SPRINGER PROTOCOLS HANDBOOKS

For further volumes:
http://www.springer.com/series/8623

Springer Protocols Handbooks collects a diverse range of step-by-step laboratory methods and protocols from across the life and biomedical sciences. Each protocol is provided in the Springer Protocol format: readily-reproducible in a step-by-step fashion. Each protocol opens with an introductory overview, a list of the materials and reagents needed to complete the experiment, and is followed by a detailed procedure supported by a helpful notes section offering tips and tricks of the trade as well as troubleshooting advice. With a focus on large comprehensive protocol collections and an international authorship, Springer Protocols Handbooks are a valuable addition to the laboratory.

Analysis of Membrane Lipids

Edited by

Rajendra Prasad

Amity Institute of Biotechnology and Amity Institute of Integrative Sciences and Health, Amity University Haryana, Gurgaon, Haryana, India

Ashutosh Singh

Department of Biochemistry, University of Lucknow, Lucknow, Uttar Pradesh, India

Springer

Editors
Rajendra Prasad
Amity Institute of Biotechnology and
Amity Institute of Integrative Sciences
and Health
Amity University Haryana
Gurgaon, Haryana, India

Ashutosh Singh
Department of Biochemistry
University of Lucknow
Lucknow, Uttar Pradesh, India

ISSN 1949-2448 ISSN 1949-2456 (electronic)
Springer Protocols Handbooks
ISBN 978-1-0716-0630-8 ISBN 978-1-0716-0631-5 (eBook)
https://doi.org/10.1007/978-1-0716-0631-5

Cover illustration: The crystal structure of green fluorescent protein (GFP) displaying a β-barrel structure with the chromophore (highlighted in orange) located in the core of the protein. GFP has been most widely used in the context of studying lateral diffusion of membrane proteins.

This Springer imprint is published by the registered company Springer Science+Business Media, LLC, part of Springer Nature.
The registered company address is: 1 New York Plaza, New York, NY 10004, U.S.A.

Preface

The first edition of *Manual on Membrane Lipids* was published back in 1996. Various techniques and protocols described in the first edition remain relevant to membrane lipid analysis to date. However in recent years, several major cutting-edge technological breakthroughs have changed the manner in which we can analyze lipids and membrane properties. For example, improved lipid extraction and purification protocols now allow higher extraction efficiency for both polar and non-polar lipids; the last decade has seen the era of "lipidomics," which is mass spectrometry-based characterization of lipids, allowing robust and accurate quantification methods using advanced techniques like electrospray ionization tandem mass spectrometry (ESI-MS/MS), gas chromatography mass spectrometry (GCMS), and nuclear magnetic resonance spectroscopy (NMR), among others; development of fluorescence-based techniques like fluorescence recovery after photobleaching (FRAP), time-resolved fluorescence spectroscopy (TRFS), and various imaging techniques now allows better visualization and understanding of membrane components and their properties. This second edition serves as a reference lipid recipe book which provides a timely overview of the groundbreaking advances made in analytical tools in the last two decades and presents methodological approaches to study high-throughput lipidomics, lipid–protein interactions, signaling pathways, regulation of lipid metabolism, functions in modulating immune systems, and diseases, among others. The lucid description preceded by updated background will make it very user friendly which every researcher and teacher would like to use it as a bench-top book. Leading experts in the field have contributed to this user-friendly bench-top manual which offers an ideal reference guide for membrane biologists, researchers, graduate and undergraduate students, clinicians, and mycologists. This book was a yearlong project. Therefore, we must thank all the authors and Springer team for their patience and support toward successful completion of this project. In the end, we would like to say that apart from being a reference book for those related in lipid research, it is extremely useful and most suited for biochemistry, microbiology, and biotechnology teaching.

Gurgaon, Haryana, India — ***Rajendra Prasad***
Lucknow, Uttar Pradesh, India — ***Ashutosh Singh***

Contents

Contributors

KHUSHBOO ARYA • *Department of Biochemistry, University of Lucknow, Lucknow, Uttar Pradesh, India*
ATANU BANERJEE • *Amity Institute of Integrative Science and Health and Amity Institute of Biotechnology, Amity University Haryana, Gurgaon, Haryana, India*
ELIANA BARRETO BERGTER • *Department of General Microbiology, Institute of Microbiology Paulo de Góes, Federal University of Rio de Janeiro (UFRJ), Rio de Janeiro, Brazil*
MARIANA COLLODETTI BERNARDINO • *Department of General Microbiology, Institute of Microbiology Paulo de Góes, Federal University of Rio de Janeiro (UFRJ), Rio de Janeiro, Brazil*
NITIN BHARDWAJ • *Department of Zoology and Environmental Science, Gurukul Kangri University, Haridwar, India*
TEJAS BOUKLAS • *Department of Biological Sciences, State University of New York College at Old Westbury, Old Westbury, NY, USA*
AMITABHA CHATTOPADHYAY • *CSIR-Centre for Cellular and Molecular Biology, Hyderabad, India*
UJJAINI DASGUPTA • *Laboratory of Sphingolipid Biology, Amity Institute of Integrative Sciences and Health, Amity University Haryana, Gurgaon, Haryana, India*
MAURIZIO DEL POETA • *Department of Molecular Genetics and Microbiology, Stony Brook University, Stony Brook, NY, USA; Veterans Administration Medical Center, Northport, NY, USA; Department of Medicine, Division of Infectious Diseases, Stony Brook University, Stony Brook, NY, USA*
KATHRIN M. ENGEL • *Faculty of Medicine, Institute for Medical Physics and Biophysics, Leipzig University, Leipzig, Germany*
AMIR M. FARNOUD • *Department of Chemical and Biomolecular Engineering, Ohio University, Athens, OH, USA*
BETTINA C. FRIES • *Department of Molecular Genetics and Microbiology, Stony Brook University, Stony Brook, NY, USA; Veterans Administration Medical Center, Northport, NY, USA; Department of Medicine, Division of Infectious Diseases, Stony Brook University, Stony Brook, NY, USA*
ASHOK KUMAR • *Amity Institute of Integrative Science and Health and Amity Institute of Biotechnology, Amity University Haryana, Gurgaon, Haryana, India*
JENNY LEOPOLD • *Faculty of Medicine, Institute for Medical Physics and Biophysics, Leipzig University, Leipzig, Germany*
NIHAL MEDATWAL • *Regional Centre for Biotechnology, NCR Biotech Science Cluster, Faridabad, Haryana, India; Manipal Academy of Higher Education, Manipal, Karnataka, India*
SANGHAMITRA MISHRA • *MedGenome Labs Ltd., Bangalore, Karnataka, India*
MANSA MUNSHI • *Department of Molecular Genetics and Microbiology, Stony Brook University, Stony Brook, NY, USA*
RODRIGO ROLLIN PINHEIRO • *Department of General Microbiology, Institute of Microbiology Paulo de Góes, Federal University of Rio de Janeiro (UFRJ), Rio de Janeiro, Brazil*
YULIA POPKOVA • *Faculty of Medicine, Institute for Medical Physics and Biophysics, Leipzig University, Leipzig, Germany*

RAJENDRA PRASAD • *Amity Institute of Integrative Science and Health and Amity Institute of Biotechnology, Amity University Haryana, Gurgaon, Haryana, India*
RAM RAJASEKHARAN • *CSIR-Central Food Technological Research Institute, Mysore, India*
PARIJAT SARKAR • *CSIR-Centre for Cellular and Molecular Biology, Hyderabad, India*
JÜRGEN SCHILLER • *Faculty of Medicine, Institute for Medical Physics and Biophysics, Leipzig University, Leipzig, Germany*
SOBHAN SEN • *Spectroscopy Laboratory, School of Physical Sciences, Jawaharlal Nehru University, New Delhi, India*
SUDHANSHU SHUKLA • *School of Medicine, Case Western Reserve University, Cleveland, OH, USA*
HIM SHWETA • *Spectroscopy Laboratory, School of Physical Sciences, Jawaharlal Nehru University, New Delhi, India*
ASHUTOSH SINGH • *Department of Biochemistry, University of Lucknow, Lucknow, Uttar Pradesh, India*
MOIRANGTHEM KIRAN SINGH • *Spectroscopy Laboratory, School of Physical Sciences, Jawaharlal Nehru University, New Delhi, India*
MALATHI SRINIVASAN • *CSIR-Central Food Technological Research Institute, Mysore, India*
SANA AKHTAR USMANI • *Department of Biochemistry, University of Lucknow, Lucknow, Uttar Pradesh, India*
AMID VAHEDI • *Department of Chemical and Biomolecular Engineering, Ohio University, Athens, OH, USA*
SACHIN DEV VERMA • *Department of Chemistry, Indian Institute of Science Education and Research Bhopal, Bhopal, Madhya Pradesh, India*
ILYA VINOGRADOV • *Department of Chemistry, University of California, Irvine, Irvine, CA, USA*

Chapter 1

Background of Membrane Lipids

Ashok Kumar, Atanu Banerjee, Ashutosh Singh, and Rajendra Prasad

Abstract

Lipids are a unique group of molecules that universally exist in both prokaryotes and eukaryotes; however, they were least investigated biomolecules owing to their water-insoluble nature. However, this scenario has been changing in the last few decades of intensive research which unraveled diverse roles played by them in a wide variety of biological processes in all spectrum of life. Notwithstanding a common footprint of lipids that exists in most organisms, there are specific lipid molecules, which are characteristic of a system. Coinciding with the development of separation and high-throughput analytical tools, we are able to detect minor lipids which otherwise remained undetected. We now know that each type of phosphoglycerides or sphingolipids is enriched with a host of molecular species imparting additional dynamism to lipid composition. These lipid changes regulate membrane homeostasis, which in turn affects the physiological functions. This chapter provides a background of lipids that are present in biological systems. Since there exists a vast amount of literature on lipid metabolism of various organisms, we will only limit our discussion to yeast systems.

Keywords Lipids, Membrane, Functions

1 Introduction

Lipids are well known as a source of biofuel and as membrane structural components; however, these sparingly water-soluble molecules have recently come to prominence due to the discovery of their multidimensional roles extending from human diseases and infections caused by microbial and fungal organisms [1]. This has fueled greater momentum in lipid research, well supported by the advancement of analytical tools including high-throughput mass spectrometry-based advanced techniques presenting high resolution and detection of lipids which otherwise remained undetected due to their lower content [2]. Among eukaryotic models for unveiling multiple roles of lipids, budding yeast *Saccharomyces cereviseae* has been a preferred model. The genetic flexibility provided by this haploid yeast has enabled us to unknot various facets of lipids. As a result, a fairly good knowledge concerning the spectrum of lipid composition, and regulatory circuitry governing

Rajendra Prasad and Ashutosh Singh (eds.), *Analysis of Membrane Lipids*, Springer Protocols Handbooks,
https://doi.org/10.1007/978-1-0716-0631-5_1,

lipid homeostasis, is available [3, 4]. The lipid metabolic pathways are generally conserved, however, with the advancement of analytical tools; newer roles of lipids impacting morbid physiology are emerging [5]. Now, lipids with manifold roles in pathophysiology acquire a prominent place among the research community.

Apart from bulk lipids like triacylglycerols, most of the lipids belonging to different classes, including phosphoglycerides (PGLs), sphingolipids (SPHs), and sterols, which are in lesser abundance, are predominantly structural component of the cell membrane [6]. Hence, imbalances in membrane lipid homeostasis often have repercussions on cellular functions. While the whole-cell lipidomics has provided a good perspective of lipids, the presence of different classes of membrane lipids demands that pure membranes are subjected to intense lipids analysis. For this, not only pure preparations of cell membranes are prerequisites but also require precision analytical tools for analyzing rare amounts of different classes of membrane lipids including the landscape of their molecular species [7–9]. The book is devoted to describe select analytical methods that provide insight into membrane lipid dynamics governing functions.

2 Classes of Lipids

The distribution of various lipid classes among various organisms is certainly not arbitrary but a consequence of their evolutionary adaption. It is not surprising that some lipid classes are more abundant in certain organisms compared to the others. For instance, while major PGLs occur in metazoans and lower eukaryotes, prokaryotes do not possess some of the major mammalian PGLs. SPHs, which are present in mammalian cells, do not universally exist in bacteria and fungi. Sterols, which are absent in prokaryotes, are a typical feature of eukaryotic cells where the most maligned molecule cholesterol is typical to mammalian cells, replaced by ergosterol in fungi and phytosterol in plants. It is quite interesting how different organisms have utilized specific lipids to maintain the lipid homeostasis of the membrane. Of course, specific lipid biosynthetic pathways regulate the differences in classes of lipids present in any organism [10].

There are eight well-identified different groups of lipids that contribute toward the overall lipid pool of various biological systems (Fig. 1). LIPID MAPS consortium, which is a recognized body for maintaining the lipid resource databases, has further classified these eight groups into 86 classes (www.lipidmaps.org). The nomenclature of various lipid species based on the guidelines from the International Union of Pure and Applied Chemists and the International Union of Biochemistry and Molecular Biology

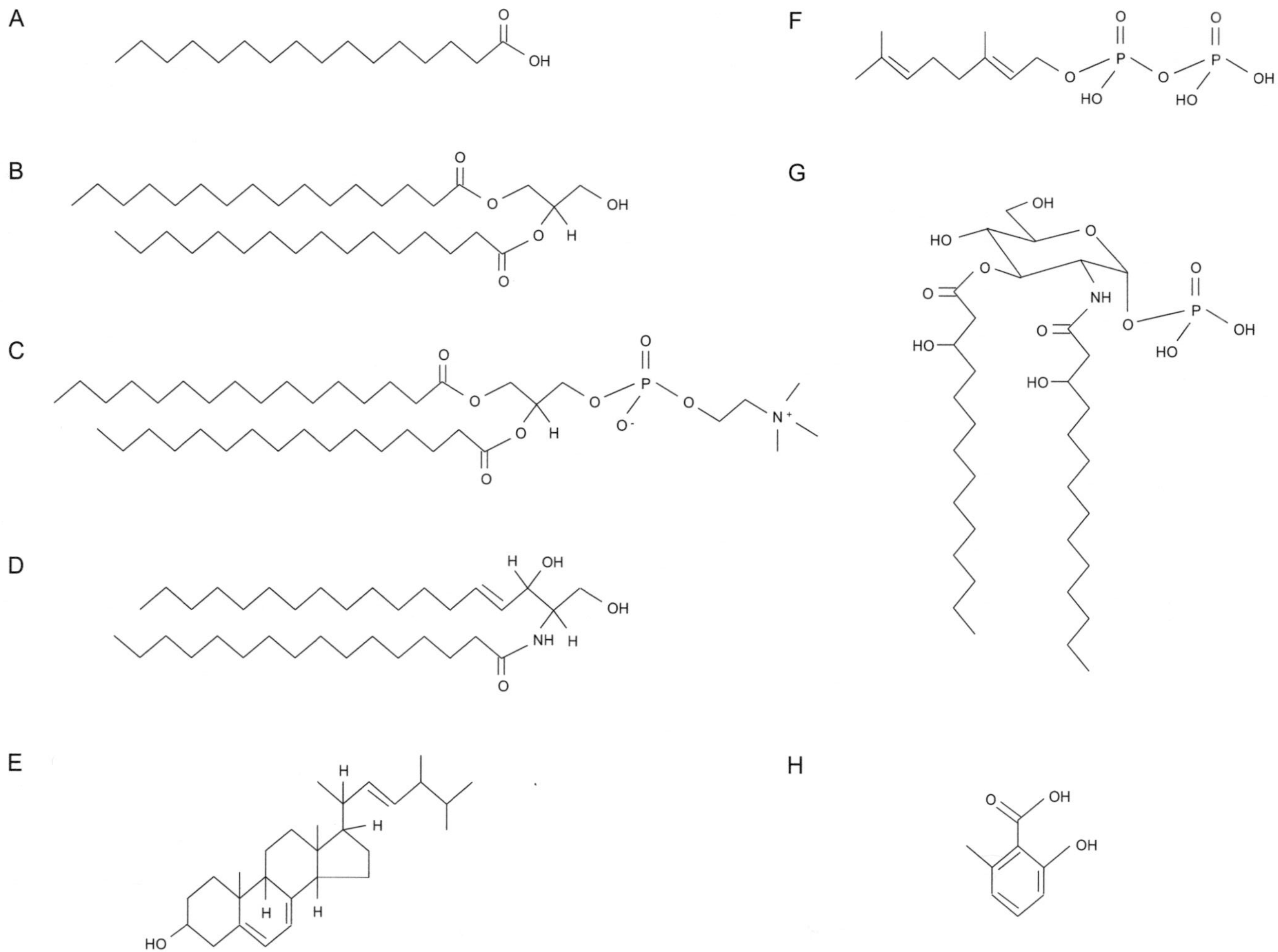

Fig. 1 Representative structures of eight classes of lipids. (**a**) palmitic acid; (**b**) 1,2-dihexadecanoyl-rac-glycerol; (**c**) dipalmitoyl phosphatidylcholine; (**d**) *N*-hexadecanoyl-D-erythro-Sphingosine C16; (**e**) ergosterol; (**f**) geranyl pyrophosphate; (**g**) Lipid-X; (**h**) 6-Methylsalicylic acid. These structures have been adopted from www.lipidmaps.org

(IUPAC-IUBMB) Commission on Biochemical Nomenclature is most widely accepted (http://www.chem.qmul.ac.uk/iupac/).

Each of these groups presents with a unique class-specific structure that is identified by their characteristic head groups or backbones [11]. These groups are named along with few common examples:

1. Fatty acyls (FA): A carbon backbone chain linked to a carboxylic acid at C1 position. The subclasses include straight-chain FA, eicosanoids, fatty alcohols, fatty esters, and fatty amides. An example of 16 carbon FA, palmitic acid or hexadecanoic acid [C16:0-FA], is shown in Fig. 1a.
2. Glycerolipids (GL): A glycerol backbone ester linked to FA at sn-1 and/or sn-2 and/or sn-3 position. The subclasses include monoacylglycerols (MAG), diacylglycerols (DAG), and triacylglycerols (TAG). An example of DAG, 1,2-dihexadecanoyl-rac-glycerol [DG(16:0/16:0/0:0)[rac]], is shown in Fig. 1b.
3. Glycerophospholipids (GPL): A glycerol backbone is ester linked to FA at the sn-1 and sn-2 with a polar head group at the sn-3 position. The subclasses are based on the polar head group. These are glycerophosphocholines (or phosphatidylcholine, PC), glycerophosphoethanolamines (or phosphatidylethanolamine, PE), glycerophosphoserines (or phosphatidylserine, PS), glycerophosphoglycerols, glycerophosphoglycerophosphates, glycerophosphoinositols (or phosphatidylinositol, PI), glycerophosphoglycerophosphoglycerols. An example of PC, 1,2-dihexadecanoyl-sn-glycero-3-phosphocholine or dipalmitoyl phosphatidylcholine [PC (16:0/16:0)], is shown in Fig. 1c.
4. SPHs: A long-chain carbon base (LCB, sphingoid base) with amine group at C2 position. The subclasses include sphingoid bases, ceramides, phosphosphingolipids, neutral glycosphingolipids, and acidic glycosphingolipids. An example of a ceramide structure, N-hexadecanoyl-D-erythro-Sphingosine C16 or N-[(1S,2R,3E)-2-hydroxy-1-(hydroxymethyl)-3-pentadecen-1-yl]-hexadecanamide Cer(d16:1/16:0)], is shown in Fig. 1d.
5. Sterol lipids: The sterol lipids have a characteristic cyclopentanoperhydrophenanthrene ring arrangement. The subclasses include sterols, sterol derivatives, steroids, and bile acids/derivatives. An example of an fungal sterol structure, ergosta-5,7,22E-trien-3β-ol or ergosterol, is shown in Fig. 1e.
6. Prenol lipids: The prenol lipids are synthesized from isoprene subunits. The subclasses include isoprenoids, quinines/hydroquinones, and polyprenols. An example of isoprenoid, geranyl pyrophosphate, is shown in Fig. 1f.

7. Saccharolipids: The saccharolipids are complex sugars containing structures. The subclasses include acylaminosugars, acylaminosugar glycans, acyltrehaloses, and acyltrehalose glycans. An example of an acylaminosugar, 2,3-bis-(3R-hydroxy-tetradecanoyl)-αD-glucosamine-1-phosphate commonly known as Lipid-X, is shown in Fig. 1g.
8. Polyketides: These ringed structures are subcharacterized into macrolide polyketides, aromatic polyketides, and non-ribosomal peptide/polyketide hybrids. An example of a monocyclic aromatic polyketide structure, 6-Methylsalicylic acid, is shown in Fig. 1h.

The diversity in the biological systems present in nature is well reflected in various lipid structures. This diversity in lipid structure depends on the structural characteristics, for example, in the case of phospholipids, it depends upon the nature of the lipid backbone, head group, and FA attached. Further diversification can occur from differences in FA chain lengths, hydroxylation, degree of unsaturation, and their position (viz. *sn*-1, *sn*-2, *sn*-3) [11, 12]. Similarly, other classes also have specific differences. This implies that a cell can manipulate the lipid biosynthetic pathways to synthesize specific lipid structures that may be required to maintain optimum homeostasis within the membrane. Thus, it becomes a pressing need to understand what are the different lipid structures present in a cell, what are their levels, how do different lipids interact with each other as well as other biomolecules like carbohydrates and proteins, and what are their roles.

In this regard, the lipid compositions of many organisms including yeasts have been determined. In budding yeasts like *S. cereviseae* and *Candida* spp., detailed lipid maps include as much as 21 lipid classes covering over 500 lipid species [3, 13–15]. An example of the lipid profile of *Candida albicans* is shown in Fig. 2. Evidently, PC, PE, MIPC (mannosylinositolphosphorylceramide), and SE (steryl esters) are among the most abundant lipids in *C. albicans* (Fig. 2a). However, the level of each PGL class is determined by the composition of PGL molecular species present therein. For example, an analysis of 35 different molecular species of PC revealed that very long-chain poly-unsaturated PC species 34:3, 34:4, 36:5, 36:4, 36:3, and 36:2 are abundant in *C. albicans* (Fig. 2b).

The remarkable lipid diversity is quite evident when we look at the respective lipid profiles of various organelles of a cell [16–18]. For example, in *S. cerevisiae*, the plasma membrane (PM) shows an enrichment of SPHs and sterols where the PGLs are relatively in low abundance. Both vacuolar (V) and endoplasmic reticulum (ER) membranes show a high content of PC, sterols, and SPHs. On the other hand, mitochondrial membranes show a completely different lipid profile where PC, PE, PI, and CL are

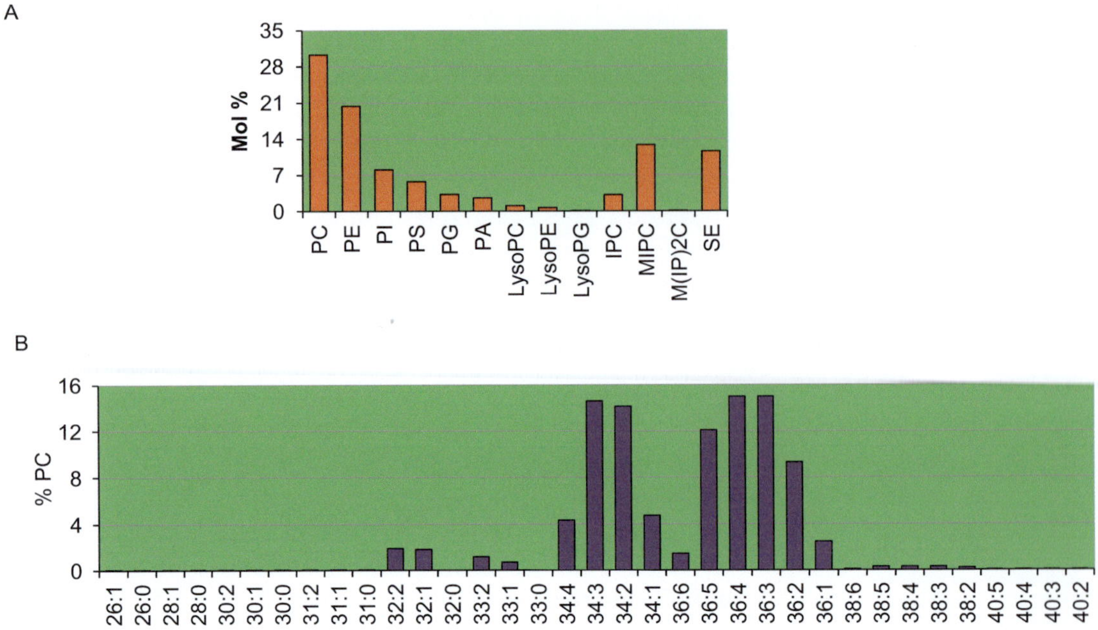

Fig. 2 Lipidome of *Candida albicans*. (**a**) a lipid class composition profile of *C. albicans*; (**b**) relative abundance of various PC species, where PC species are represented as "total number of carbon atoms in the FA chains: total number of double bonds in the FA chains"

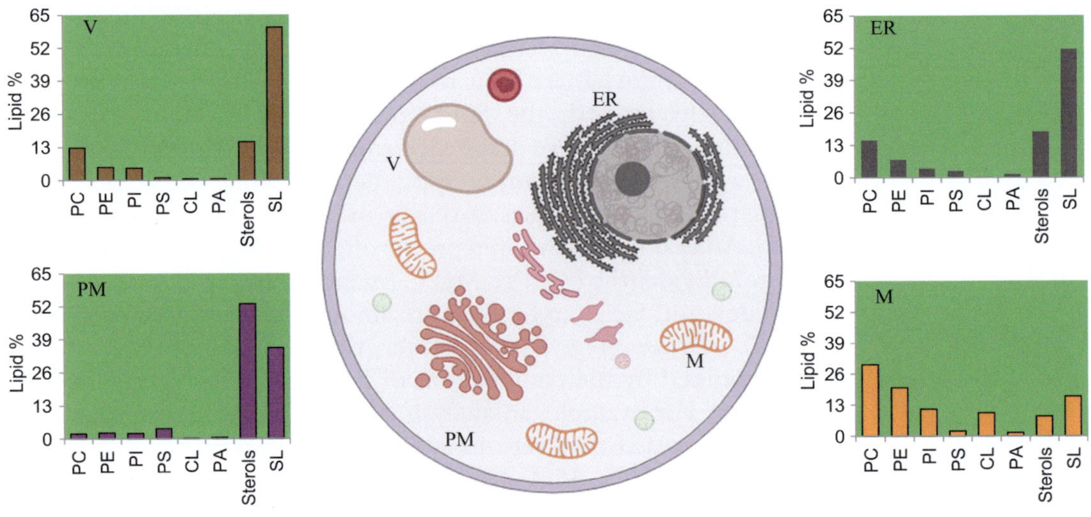

Fig. 3 Membrane lipid compositions of various cellular organelles of a *S. cerevisiae*. Here "V" represents the vacuolar membrane, "PM" represents the plasma membrane, "ER" represents the endoplasmic reticulum, and "M" represents the mitochondrial membrane, "CL" and "SL" represent cardiolipin and sphingolipid, respectively

more abundant compared to their levels in PM, ER, and vacuolar membranes (Fig. 3). The differences in specific lipid compositions of these membranes are evidently necessary to maintain their lipid homeostasis or membrane environment to perform specific physiological functions.

3 Unique Functional Constituents of Lipids

Notwithstanding the common lipids, which are present in different organisms, many are bestowed with a unique variant of lipids [12]. For instance, mammalian cell's typical distinction from lower eukaryotes is reflected in SPH composition where its lipid backbones are characteristically derivatized via linkage at the 1-hydroxy position of either complex carbohydrates or other polar species such as PC or PE. Contrasting this, typical fungi where SPHs are equally relevant to cellular physiology, their SPH's 1-hydroxy position is derivatized with phosphorylated inositols. In another example, glucosylceramide (GlcCer) structure in mammals is composed of d18:1 sphingoid backbone which is N-linked to fatty acyls of varied chain lengths. However, in fungi, the GlcCer or galactosylceramide structure is composed of either d18:1, d18:2, or d19:2 (most abundant structure) backbones which are N-linked to α-hydroxylated fatty acyls, mostly C18:0h [19]. Cholesterol is again one of the typical lipids which constitutes a major part of mammalian lipids as compared to ergosterol in fungi. Some fungi also have typical sterol intermediates like acylated sterols and glucosylated sterols [20]. Some of the typical lipids are depicted in Fig. 1.

4 Diverse Membrane Lipid Composition, Microdomain, and Asymmetry

Typical membrane contains PGLs, SPHs, and sterols albeit in different proportions [10]. Notably, this unique membrane composition is what makes a membrane distinct from other membranes. Superimposed with this, the plasma membrane is remarkably asymmetric, with different classes of lipids occupying the outer and cytosolic leaflets of the membrane bilayer. For instance, aminolipids such as PE and PS are typically located within the cytoplasmic leaflet of bilayer while PC is normally equally distributed between the two leaflets. The typical asymmetrical distribution and composition are well maintained by a host of flippase, floppase, and scramblase enzymes, which control trans-bilayer moment and distribution of lipids [21, 22]. Apart from this distinction, membranes also harbor microdomains (membrane rafts) with specific lipid constituents like sterols and SPHs. The microdomain serves as a platform for a variety of cellular signaling and sites for select proteins' localization and function [23]. There are great functional implications associated with lipid diversity that prevails in membranes in terms of their composition and asymmetrical distribution (Table 1).

Table 1
Select roles of yeast/fungal lipids

S. No.	Phenomenon	Lipid species	References
1	Cellular ageing		[24]
	(a) Replicative life span	Sphingosine and/or ceramide species	
	(a) Chronological life span	Mannosyl-di-inositol-phosphorylceramide (M(IP)$_2$C), ceramides, mitochondrial membrane phospholipids, Tri-acyl glycerol (TAG)	
2.	Endomembrane system and associated cellular trafficking	Glycerophospholipids (GPLs), sphingolipids (SPHs), and ergosterol	[25]
3.	Physicochemical properties of the membrane	Glycerophospholipids, ergosterol	[26]
4.	Stress response	Sphingolipids	[27, 28]
5.	Cell cycle control	Sphingolipids	[29]
6.	Drug resistance	Ergosterol	[30]
7.	Biofilm formation	Phospholipids, ergosterol, and sphingolipids	[31]
8.	Virulence and pathogenesis	Sphingolipids species such as glucosylceramide, sterylglucosides, and ergosterol	[20]
9.	Cellular signaling	Sphingolipids, farnesol, and oxylipins	[32]
10.	Modulation of host immune responses	Eicosanoids like prostaglandins (PG)	[33]

5 Importance of Analytical Techniques to Understand Membranes and Lipid Structure/Function

As discussed above, lipids are classified into eight broad groups, which are subdivided into several hundred lipid classes, which are in turn composed of several thousand molecular lipid species (www.lipidmaps.org). Each molecular lipid species is not only of structural value but also plays several important roles within the cell. This demands an accurate determination of lipid compositions at the molecular species level. Earlier, these lipid analyses were limited to techniques like thin-layer chromatography (TLC), gas chromatography (GC), gas-liquid chromatography (GLC), and high-performance liquid chromatography (HPLC) that enabled us to determine the lipid class compositions in a semi-quantitative manner [34–36]. These techniques mostly relied on the availability of appropriate standards. While analysis on TLC covers most lipid classes, GC/GLC can analyze derivatized FA and sterol intermediates. HPLC is well suited for sterols, FA, PGLs, DAGs, and TAGs. Additionally, radioactive labeling of lipid precursors followed by resolution on TLC was also a quite common method to monitor the lipid biosynthetic pathways. Apart from iodine staining, the TLC-resolved lipids can also be visualized on the basis of specific coloring reagents [35]. For example, choline containing lipids like PC, lyso-PC (LPC), and SM give an orange-red spot when stained with bismuth nitrate + KI solution (Dragendorff reagent). Similarly, amino group containing PGLs like PE, PS, lyso-PE (LPE), lyso-PS (LPS) stain as red-violet spots in ninhydrin in butanol solution.

However, a major breakthrough was achieved when the mass spectrometry was developed as a tool to determine the lipid compositions in the 1990s. Soft ionization techniques like electrospray ionization (ESI) coupled with tandem mass spectrometers (MS/MS), atmospheric pressure chemical ionization (APCI), and later matrix-assisted laser desorption ionization (MALDI) were most commonly used. ESI and MADLI ionization cover almost all lipid classes including the glycolipids [37–40]. However, APCI is preferred for lipids with high polarity. However, both ESI and MADLI suffer from the problem of ion suppression; therefore, a prior HPLC/TLC separation of the crude extract is much recommended. Mass spectrometers coupled with an HPLC or gas chromatography have become a modern-day tool to analyze lipids. By the year 2000, mass spectrometry became much advanced, and with the evolution of hybrid mass spectrometers like the Orbitrap, accurate mass measurements of molecular lipid species became possible. The lipid structures and quantification can also be performed using nuclear magnetic resonance spectroscopy (NMR). For example, ^{31}P-NMR can be used to detect the class composition of PLs, and ^{1}H-NMR can potentially analyze the structures of most lipids; however, the sensitivity of detection is a major challenge with the analysis of lipids using NMR [41]. In addition to this, several fluorescent labeling-based methods have also been developed that can directly indicate the nature of membrane environment. For example, a time-resolved fluorescence analysis and steady-state measurements can be done of any biological or artificial membrane using various fluorescent probes, like fluorescamine-, rhodamine-, or nitrobenzoxadiazole (NBD)-labeled lipids [42–44]. In addition, several indirect spectrophotometry-based methods have been developed to determine the levels of sterols, specifically ergosterol and dehydroergosterol. Obviously, the choice of lipid quantification method depends on the utility and availability.

Acknowledgments

We thank grants to RP from DBT No. BT/01/CEIB/10/III/02, BT/PR7392/MED/29/652/2012 and BT/PR14117/BRB/

10/1420/2015. We thank financial assistance to AS from ICMR No. 52/08/2019-BMS and University of Lucknow, Lucknow. AS thanks Amity University, Haryana for inviting for a mini sabbatical and support therein. AB acknowledges the financial support from SERB Grant no. SRG/2019/000514.

Financial and Competing Interest Disclosure: *There is no financial and competing interest.*

Contribution to the Manuscript: *RP, AB, AK, and AS wrote the manuscript.*

References

1. Chen Z, Wang L, Qiu S, Ge S (2018) Determination of microalgal lipid content and fatty acid for biofuel production. Biomed Res Int 2018:1503126
2. Yang K, Cheng H, Gross RW, Han X (2009) Automated lipid identification and quantification by multidimensional mass spectrometry-based shotgun lipidomics. Anal Chem 81:4356–4368
3. Ejsing CS, Sampaio JL, Surendranath V, Duchoslav E, Ekroos K, Klemm RW, Simons K, Shevchenko A (2009) Global analysis of the yeast lipidome by quantitative shotgun mass spectrometry. Proc Natl Acad Sci U S A 106(7):2136–2141
4. Casanovas A, Sprenger RR, Tarasov K, Ruckerbauer DE, Hannibal-Bach HK, Zanghellini J, Jensen ON, Ejsing CS (2015) Quantitative analysis of proteome and lipidome dynamics reveals functional regulation of global lipid metabolism. Chem Biol 22(3):412–425
5. Han X (2016) Lipidomics for studying metabolism. Nat Rev Endocrinol 12(11):668–679
6. Harayama T, Riezman H (2018) Understanding the diversity of membrane lipid composition. Nat Rev Mol Cell Biol 19(5):281–296
7. Singh A, Del Poeta M (2016) Sphingolipidomics: an important mechanistic tool for studying fungal pathogens. Front Microbiol 7:501
8. de Kroon AIPM (2017) Lipidomics in research on yeast membrane lipid homeostasis. Biochim Biophys Acta Mol Cell Biol Lipids 1862 (8):797–799
9. Singh A, Khandelwal NK, Prasad R (2019) Lipidomics approaches: applied to the study of pathogenesis in *Candida*species. Prog Mol Subcell Biol. 58:195–215
10. van Meer G, Voelker DR, Feigenson GW (2008) Membrane lipids: where they are and how they behave. Nat Rev Mol Cell Biol 9 (2):112–124
11. Fahy E, Cotter D, Sud M, Subramaniam S (2011) Lipid classification, structures and tools. Biochim Biophys Acta 1811 (11):637–647
12. Wenk MR (2010) Lipidomics: new tools and applications. Cell 143(6):888–895
13. Singh A, Prasad T, Kapoor K, Mandal A, Roth M, Welti R, Prasad R (2010) Phospholipidome of *Candida*: each species of *Candida* has distinctive phospholipid molecular species. OMICS 14(6):665–677
14. Singh A, Prasad R (2011) Comparative lipidomics of azole sensitive and resistant clinical isolates of *Candida albicans* reveals unexpected diversity in molecular lipid imprints. PLoS One 6(4):e19266
15. Singh A, Yadav V, Prasad R (2012) Comparative lipidomics in clinical isolates of *Candida albicans* reveal crosstalk between mitochondria, cell wall integrity and azole resistance. PLoS One 7(6):e39812
16. Zinser E, Daum G (1995) Isolation and biochemical characterization of organelles from the yeast, *Saccharomyces cerevisiae*. Yeast 11 (6):493–536
17. Horvath SE, Daum G (2013) Lipids of mitochondria. Prog Lipid Res 52(4):590–614
18. Brügger B (2014) Lipidomics: analysis of the lipid composition of cells and subcellular organelles by electrospray ionization mass spectrometry. Annu Rev Biochem 83:79–98
19. Del Poeta M, Nimrichter L, Rodrigues ML, Luberto C (2014) Synthesis and biological properties of fungal glucosylceramide. PLoS Pathog 10(1):e1003832
20. Rella A, Farnoud AM, Del Poeta M (2016) Plasma membrane lipids and their role in fungal virulence. Prog Lipid Res 61:63–72
21. Ikeda M, Kihara A, Igarashi Y (2006) Lipid asymmetry of the eukaryotic plasma

membrane: functions and related enzymes. Biol Pharm Bull 29(8):1542–1546

22. Kobayashi T, Menon AK (2018) Transbilayer lipid asymmetry. Curr Biol 28(8):R386–R391
23. Farnoud AM, Toledo AM, Konopka JB, Del Poeta M, London E (2015) Raft-like membrane domains in pathogenic microorganisms. Curr Top Membr 75:233–268
24. Mitrofanova D, Dakik P, McAuley M, Medkour Y, Mohammad K, Titorenko VI (2018) Lipid metabolism and transport define longevity of the yeast *Saccharomyces cerevisiae*. Front Biosci (Landmark Ed) 23:1166–1194
25. Lippincott-Schwartz J, Phair RD (2010) Lipids and cholesterol as regulators of traffic in the endomembrane system. Annu Rev Biophys 39:559–578
26. Renne MF, de Kroon AIPM (2018) The role of phospholipid molecular species in determining the physical properties of yeast membranes. FEBS Lett 592:1330–1345
27. Patton JL, Srinivasan B, Dickson RC, Lester RL (1992) Phenotypes of sphingolipid-dependent strains of *Saccharomyces cerevisiae*. J Bacteriol 174:7180–7184
28. Jenkins GM, Richards A, Wahl T et al (1997) Involvement of yeast sphingolipids in the heat stress response of *Saccharomyces cerevisiae*. J Biol Chem 272:32566–32572
29. Cheng J, Park TS, Fischl AS, Ye XS (2001) Cell cycle progression and cell polarity require sphingolipid biosynthesis in *Aspergillus nidulans*. Mol Cell Biol 21:6198–6209
30. Prasad R, Banerjee A, Shah AH (2017) Resistance to antifungal therapies. Essays Biochem 61:157–166
31. Alim D, Sircaik S, Panwar LS (2018) The significance of lipids to biofilm formation in *Candida albicans*: an emerging perspective. J Fungi (Basel) 4(4):E140
32. Singh A, Del Poeta M (2011) Lipid signalling in pathogenic fungi. Cell Microbiol 13:177–185
33. Noverr MC, Phare SM, Toews GB et al (2001) Pathogenic yeasts *Cryptococcus neoformans* and *Candida albicans* produce immunomodulatory prostaglandins. Infect Immun 69:2957–2963
34. Nissen HP, Kreysel HW (1990) The use of HPLC for the determination of lipids in biological materials. Chromatographia 30 (11–12):686–690
35. Fuchs B, Süss R, Teuber K, Eibisch M, Schiller J (2011) Lipid analysis by thin-layer chromatography--a review of the current state. J Chromatogr A 1218(19):2754–2774
36. Fisk HL, West AL, Childs CE, Burdge GC, Calder PC (2014) The use of gas chromatography to analyze compositional changes of fatty acids in rat liver tissue during pregnancy. J Vis Exp. 85:51445
37. Han X, Gross RW (1994) Electrospray ionization mass spectroscopic analysis of human erythrocyte plasma membrane phospholipids. Proc Natl Acad Sci U S A 91 (22):10635–10639
38. Jones EE, Dworski S, Canals D, Casas J, Fabrias G, Schoenling D, Levade T, Denlinger C, Hannun YA, Medin JA, Drake RR (2014) On-tissue localization of ceramides and other sphingolipids by MALDI mass spectrometry imaging. Anal Chem 86 (16):8303–8311
39. Shaner RL, Allegood JC, Park H, Wang E, Kelly S, Haynes CA, Sullards MC, Merrill AH Jr (2009) Quantitative analysis of sphingolipids for lipidomics using triple quadrupole and quadrupole linear ion trap mass spectrometers. J Lipid Res 50(8):1692–1707
40. Sun Q, Gu J, Stolze BR, Soldin SJ (2018) Atmospheric pressure chemical ionization is a suboptimal ionization source for steroids. Clin Chem 64(6):974–976
41. Singh A, MacKenzie A, Girnun G, Del Poeta M (2017) Analysis of sphingolipids, sterols, and phospholipids in human pathogenic *Cryptococcus*strains. J Lipid Res 58(10):2017–2036
42. Lee HC, Forte JG (1979) Asymmetric labeling of amino lipids in liposomes. Biochim Biophys Acta 554(2):375–387
43. Chazotte B (2011) Labeling membranes with fluorescent phosphatidylethanolamine. Cold Spring Harb Protoc. 2011(5). https://doi.org/10.1101/pdb.prot5621
44. Amaro M, Filipe HA, Prates Ramalho JP, Hof M, Loura LM (2016) Fluorescence of nitrobenzoxadiazole (NBD)-labeled lipids in model membranes is connected not to lipid mobility but to probe location. Phys Chem Chem Phys 18(10):7042–7054

Chapter 2

Lipid Regulation in Pathogenic Fungi

Tejas Bouklas, Mansa Munshi, Maurizio Del Poeta, and Bettina C. Fries

Abstract

Much of the current research on lipids points to their role in regulating the infectious disease process of pathogenic microorganisms. This is particularly important in the case of sterols, which vary among the major pathogenic fungi and contribute differentially to their susceptibility to antifungals. Unfortunately, studies of such pathogenic fungi are limiting, from incomplete lipid extraction and analysis. In addition to challenges posed by mutant collections of pathogenic fungi, including *Cryptococcus neoformans* and *Candida* spp., fungal characteristics, such as encapsulation or biofilm formation, further complicate studies. This chapter outlines successful modifications made to traditional methods for such fungal pathogens.

Keywords Lipids, Sterols, Azoles, Extraction, Pathogen, Fungi, *Cryptococcus*, *Candida*

1 Introduction

While lipids are widely known as energy storage molecules and cell membrane components, their role in membrane signaling has now become quite prominent. The role of lipids in microbial cells varies, but has been generally been agreed to include cellular growth, apoptosis, and differentiation [1–3]. Much of the current research also points to the role of lipids in regulating the infectious disease process of pathogenic microorganisms. This is particularly important in the case of sterols, which vary among the major pathogenic fungi and have been shown to contribute differentially to their susceptibility to antifungals, specifically azoles [4]. Unfortunately, studies of such fungi are limiting, especially for clinical strains that may resist complete lipid extraction and therefore analysis.

Since most studies on fungal sterols are done using mutant library strains of *Saccharomyces cerevisiae* [5, 6], much of the knowledge on their composition in other fungal groups, especially in pathogenic fungi such as *Candida* spp. and *Cryptococcus neoformans*, is limiting. Further *Candida* and *Cryptococcal* mutant collections pose additional challenges to lipid composition investigations

Rajendra Prasad and Ashutosh Singh (eds.), *Analysis of Membrane Lipids*, Springer Protocols Handbooks,
https://doi.org/10.1007/978-1-0716-0631-5_2,

as mutant strains may show unusually large cell bodies, cell walls, or capsules [7–9].

This chapter outlines modifications to well-established methods of lipid extraction for fungi that resist such isolation and make it difficult to analyze lipids, including inositol-containing phospholipids, phosphatidylcholine, neutral lipids, and glucosylceramide. One reason for the interest in these lipid groups is the intriguing link between phospholipids and sterols. Greater quantities of non-esterified sterol were found in azole-resistant *Candida* spp., where the phospholipid to sterol ratio was 50% less compared to azole-sensitive strains [10]. This was attributed by the group to membrane permeability differences, and not limited ergosterol synthesis. A laboratory generated mutant of *C. albicans*, showing azole and polyene resistance, had replaced ergosterol with methylated sterols, specifically 4-methylergostadiene-3-ol and 24-methylene-24,25-dihydrolanosterol, and therefore lacked an important active site for azoles [11]. Therefore, for *Candida* and potentially the other fungal pathogens, this link will need to be further elucidated.

Other potential fungal pathogens include encapsulated fungi, such as *Cryptococcus neoformans*, or biofilm-forming fungi, such as *Candida albicans*. This method may also prove useful in extracting lipids for further analysis in the more understudied fungi, such as replicatively or chronologically older fungi [7, 8], or fungi with unusually thickened cell wall or cell membranes, which may be the case for sterol mutants [12]. Importantly, this method may aid analysis of clinical strains from patients that have undergone extensive passage in hosts [13], and may resist traditional extraction methods. A proof of principle for this method is shown in Fig. 1.

2 Materials and Methods

2.1 Cell Harvest

1. Under sterile conditions, fill a 50 ml conical tube with 9 ml yeast peptone (YP) and 1 ml 20% glucose (*see* **Note 1**).
2. Add a single colony of the strain of interest to the tube.
3. Incubate the tube in a shaking incubator for 16 h at 37 °C and 150 rpm (*see* **Notes 2** and **3**).
4. After incubation, centrifuge at 1620 × *g* for 10 min at 4 °C.
5. Wash the pellet twice with phosphate-buffered saline (PBS), then resuspend in 9 ml PBS (*see* **Note 4**).
6. Count the cells on a hemocytometer using an appropriate serial dilution.
7. Aliquot 5×10^8 cells per tube.
8. Centrifuge tubes at 3000 rpm for 10 min at 4 °C.
9. Remove the supernatant carefully.

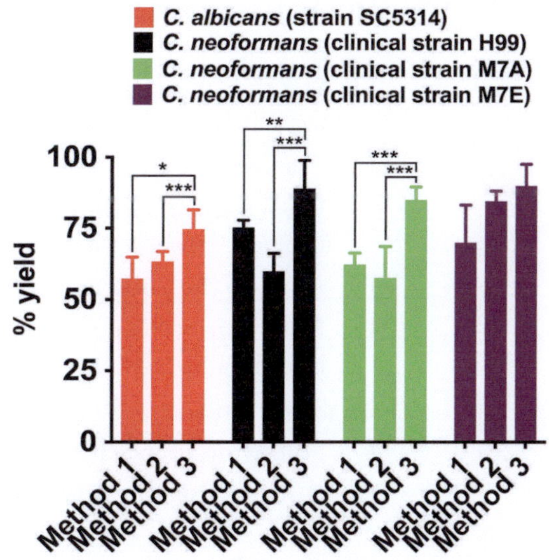

Fig. 1 Lipid extraction efficiency using outlined method for several pathogenic fungi. Method 1 ([14]) and method 2 [7] yielded fewer neutral lipids compared to the modified method presented in this chapter. The yield was higher and most significant for standard laboratory strains, *Candida albicans* SC5314, *Cryptococcus neoformans* H99, and *C. neoformans* clinical strains M7A (first isolate from patient at Montefiore Medical Center), but not strain M7E (fifth isolate from patient). Results were compared by ANOVA using Graphpad Prism 7 for Macintosh

2.2 First Lipid Extraction for Inositol-Containing Phospholipids and Phosphatidylcholine (Modified from [14, 15])

1. Pellet each sample in a glass tube and add 1.5 ml of mandala extraction buffer (150 ml distilled water, 150 ml ethanol, 50 ml diethyl ether, 10 ml pyridine, and 180 μl 14.2 N ammonium hydroxide).
2. Vortex and sonicate for 20 s each, and incubate at 60 °C in a water bath for 15 min.
3. Repeat the earlier step two more times (*see* **Note 5**).
4. Sonicate for 20 s, then centrifuge for 10 min at 3000 rpm at 4 °C.
5. Combine supernatant from no more than two tubes into a clean tube using a glass Pasteur pipette.
6. Evaporate the solvent in a SpeedVac (*see* **Note 6**).

2.3 Second Lipid Extraction for Neutral Lipids (Modified from [16])

7. Suspend a maximum of 10^8 cells for each sample in 10 ml methanol (*see* **Note 7**).
8. Add 4 g glass beads (Glaperlon 0.40–0.60 mm) to the suspension.
9. Shake in a cell disintegrator (B. Braun, Melsungen, Germany) four times for 30 s with a gap of 30 s between shakings.

10. Add approximately 20 ml chloroform to the suspension to give a ratio of 2:1 of chloroform:methanol (v/v).
11. Stir the suspension on a flat-bed stirrer at room temperature for 2 h and then filter through Whatman No. 1 filter paper.
12. Transfer the extract to a separating funnel.
13. Wash with 0.2 volumes of 0.9% sodium chloride to remove the non-lipid contaminants.
14. Aspirate the aqueous layer.
15. Evaporate the solvent of the lipid-containing, lower organic layer in a Speedvac or preferably under liquid nitrogen.
16. Store lipids at −80 °C until further analysis (*see* **Note 8**).
17. Following evaporation, add 2 ml methanol and vortex.
18. Sonicate for 20 s if necessary.

2.4 Base Hydrolysis

19. To each sample, add 0.5 ml chloroform, then 0.5 ml 0.6 M potassium hydroxide in methanol.
20. Vortex well and incubate at room temperature for 1 h.
21. Add 0.325 ml 1 M hydrochloric acid, then 0.125 ml distilled water.
22. Vortex and centrifuge at 3000 rpm for 10 min at room temperature.
23. Transfer the lower organic phase to a clean tube.
24. Evaporate the solvent in a SpeedVac.
25. At this point, a small dark brown pellet should be visible. Weigh the pellet (*see* **Note 9**).

2.5 Lipid Analysis Using Thin Layer Chromatography

26. Prepare a glass TLC tank with a chloroform:methanol:water solution (97.5:37.5:6).
27. Line tank with white chromatography paper (*see* **Note 10**), then apply a thin layer of vacuum grease around its top to ensure a good seal. Leave alone until paper is well saturated or at least 5 h but no more than 16 h (*see* **Note 11**).
28. Remove the TLC plate to spot appropriate standards approximately 1.5 cm from the bottom using an extended use 10 μl pipette (*see* **Note 12**). Place 1 μl in each lane, which is equivalent to 2.5 μg (*see* **Note 13**).
29. Resuspend the dessicated lipid from step 25 in up to 30 μl chloroform:methanol (2:1), and test 5 μl or analyze complete sample onto the TLC plate into a lane next to the standards.
30. Let the solvent evaporate in a fume hood, then place the TLC plate back into the tank.
31. Ensure that the TLC tank is tightly closed (*see* **Note 11**).

32. Let the solvent front to migrate to the top of the plate up to 1 cm from it before removing the plate.
33. Dry the TLC plate in the fume hood at room temperature before placing it in a second tank with iodine crystals to visualize the lipids (*see* **Note 14**).

2.6 Lipid Analysis Using Gas Chromatography-Mass Spectrometry

34. For GC-MS (*see* **Note 15**), derivatize the desiccated sample from step 25 using *N*,*O*-bis (trimethylsilyl) trifluoroacetamide/trimethylchlorosilane (Sigma-Aldrich).
35. Analyze the sample using a 30 m (0.25 μm) VF-5 ms column on an Agilent 7890 GC-MS (Agilent Technologies, Santa Clara, CA, USA).
36. The retention time of the ergosterol standard (Matreya LLC, PA) may be used as a reference.
37. Add cholesterol as an internal standard for these analyses prior to lipid extraction.

3 Notes

1. To avoid clumping often seen in encapsulated fungi, such as *Cryptococcus neoformans*, or biofilm forming fungi, such as *Candida albicans*, two other media are suggested:
   (a) SD media: 6.7 g yeast nitrogen base without amino acids supplemented with 20 g glucose per liter of sterile water. This media must be vacuum filtered.
   (b) SD+ media: SD media supplemented with 0.4% ethanol, 5 g ammonium sulfate, and 3.3 g sodium chloride per liter of water. This media extensively inhibits capsule formation, and also must be vacuum filtered.
2. Temperature can be altered to 22–42 °C depending on the strain used.
3. It is important to harvest cells at an exponential phase to maximize lipid extraction.
4. To avoid clumping often seen in encapsulated fungi or biofilm-forming fungi, wash cells twice with 30 mM ethylenediamine-tetraacetic acid (EDTA), and then thrice with PBS. For gentler washes or when dealing with clinical samples to ensure viability, a 1:1 EDTA:PBS wash may be used twice, followed by a PBS wash used thrice.
5. Earlier methods [17] repeat this step two times; we find an increased yield with a longer time in Mandala buffer.
6. Original reference for this method is [18].

7. This step deviates from earlier methods as it scales up for maximal yield. The original reference for this method is [17].
8. Lipids may be stored for up to 2 weeks without significant loss of yield. Longer times have not been tested.
9. Earlier methods [16] recommend that if contamination of non-lipids is present, the chloroform should be decanted and the flask rinsed with fresh chloroform. In this case, the weight of the pellet should be calculated again and subtracted from the initial weight.
10. Draw a faint line with a pencil approximately 1.5 cm from the bottom before adding to the tank.
11. Consider laying a particle board shelf on top to ensure a good seal.
12. Soy is a traditional standard. However, synthetic lipids with fatty acid compositions that are not found or of low abundance in *Candida* can be used as other standards (Avanti Polar Lipids, Alabaster, AL).
13. If suspected yields are low, or for clinical samples, up to 3 μl may be used in each lane (equivalent to 7.5 μg).
14. If visualization is difficult, other methods [17] suggest spraying the plate with resorcinol in 70% sulfuric acid, then placing it in a dry oven for 10 min to let a dark purple color develop in spots where sugars are located on the lipids.
15. The dried sample can also be analyzed by liquid chromatography instead of gas chromatography. LC-MS is particularly useful to quantitate sito and glycosylated ergosterol. A standard curve will need to be generated with sito and can be used to quantify glycosylated ergosterol.

References

1. Dickson RC, Lester RL (2002) Sphingolipid functions in *Saccharomyces cerevisiae*. Biochim Biophys Acta 1583(1):13–25. https://doi.org/10.1016/s1388-1981(02)00210-x
2. Noverr MC, Erb-Downward JR, Huffnagle GB (2003) Production of eicosanoids and other oxylipins by pathogenic eukaryotic microbes. Clin Microbiol Rev 16(3):517–533. https://doi.org/10.1128/cmr.16.3.517-533.2003
3. Hanada K (2005) Sphingolipids in infectious diseases. Jpn J Infect Dis 58(3):131–148
4. Pierce AM, Pierce HD Jr, Unrau AM, Oehlschlager AC (1978) Lipid composition and polyene antibiotic resistance of *Candida albicans* mutants. Can J Biochem 56(2):135–142
5. Brennan PJ, Losel DM (1978) Physiology of fungal lipids: selected topics. Adv Microb Physiol 17:47–179
6. Rattray JB, Schibeci A, Kidby DK (1975) Lipids of yeasts. Bacteriol Rev 39(3):197–231
7. Bouklas T, Alonso-Crisostomo L, Szekely T Jr, Diago-Navarro E, Orner EP, Smith K et al (2017) Generational distribution of a *Candida glabrata* population: resilient old cells prevail, while younger cells dominate in the vulnerable host. PLoS Pathog 13(5):e1006355. https://doi.org/10.1371/journal.ppat.1006355
8. Bouklas T, Pechuan X, Goldman DL, Edelman B, Bergman A, Fries BC (2013) Old *Cryptococcus neoformans* cells contribute to virulence in chronic cryptococcosis. MBio 4(4). https://doi.org/10.1128/mBio.00455-13

9. Trevijano-Contador N, de Oliveira HC, Garcia-Rodas R, Rossi SA, Llorente I, Zaballos A et al (2018) *Cryptococcus neoformans* can form titan-like cells in vitro in response to multiple signals. PLoS Pathog 14(5):e1007007. https://doi.org/10.1371/journal.ppat.1007007
10. Hitchcock CA, Barrett-Bee KJ, Russell NJ (1986) The lipid composition of azole-sensitive and azole-resistant strains of *Candida albicans*. J Gen Microbiol 132(9):2421–2431. https://doi.org/10.1099/00221287-132-9-2421
11. Hitchcock CA, Barrett-Bee KJ, Russell NJ (1987) The lipid composition and permeability to azole of an azole- and polyene-resistant mutant of *Candida albicans*. J Med Vet Mycol 25(1):29–37
12. Munshi MA, Gardin JM, Singh A, Luberto C, Rieger R, Bouklas T et al (2018) The role of ceramide synthases in the pathogenicity of *Cryptococcus neoformans*. Cell Rep 22(6):1392–1400. https://doi.org/10.1016/j.celrep.2018.01.035
13. Fries BC, Casadevall A (1998) Serial isolates of *Cryptococcus neoformans* from patients with AIDS differ in virulence for mice. J Infect Dis 178(6):1761–1766. https://doi.org/10.1086/314521
14. Singh A, Yadav V, Prasad R (2012) Comparative lipidomics in clinical isolates of *Candida albicans* reveal crosstalk between mitochondria, cell wall integrity and azole resistance. PLoS One 7(6):e39812. https://doi.org/10.1371/journal.pone.0039812
15. Mandala SM, Thornton RA, Frommer BR, Curotto JE, Rozdilsky W, Kurtz MB et al (1995) The discovery of australifungin, a novel inhibitor of sphinganine N-acyltransferase from *Sporormiella australis*. Producing organism, fermentation, isolation, and biological activity. J Antibiot (Tokyo) 48(5):349–356
16. Bligh EG, Dyer WJ (1959) A rapid method of total lipid extraction and purification. Can J Biochem Physiol 37(8):911–917. https://doi.org/10.1139/o59-099
17. Singh A, Qureshi A, Del Poeta M (2011) Quantitation of cellular components in *Cryptococcus neoformans* for system biology analysis. Methods Mol Biol 734:317–333. https://doi.org/10.1007/978-1-61779-086-7_16
18. Hanson BA, Lester RL (1980) The extraction of inositol-containing phospholipids and phosphatidylcholine from *Saccharomyces cerevisiae* and *Neurospora crassa*. J Lipid Res 21(3):309–315

Chapter 3

Sphingolipids: Functional and Biological Aspects in Mammals, Plants, and Fungi

Rodrigo Rollin-Pinheiro, Mariana Collodetti Bernardino, and Eliana Barreto-Bergter

Abstract

Sphingolipids are an important class of lipids found in a variety of organisms, including mammals, plants, and fungi. It is structurally diverse among them, presenting variations in the headgroups as well as in the sites of unsaturation and length of the fatty acid chain. Although structurally different, some molecules are found in mammalian, fungal, and plant cells, such as glycosylceramides. On the other hand, there are sphingolipids only found in certain organisms, such as gangliosides and sphingomyelin in mammals, 9-methyl-4,8-sphingadienine in fungi, and specific inositolphosphorylceramides in plants. A variety of methodologies are available in the literature in order to extract, purify, and identify sphingolipid structures, all of them based on the use of organic solvents, chromatographic, and spectrometric techniques. The study of sphingolipids shows to be necessary when considering the roles for biological events, such as membrane integrity and cell morphology, as well as cellular signaling, nutrient uptake, and regulation of growth. For these reasons, this chapter aims to discuss the most important aspects of sphingolipids that have been studying during the last few decades.

Keywords Sphingolipids, Glycosylceramides, Inositolphosphorylceramides, Gangliosides, Sphingomyelin, Ceramide, Long-chain base, Fatty acid

1 Introduction

Sphingolipids are defined as a class of lipid molecules composed of a sphingoid backbone linked to a fatty acid chain through an amide bond [1]. A variety of sphingolipids are found in nature, especially those from eukaryotic cells, such as mammals, plants, and fungi.

It is generally considered a conserved lipid, although crucial structural differences are observed mainly regarding the number of carbons of the fatty acid chain as well as the unsaturation level [2]. Chemical characterization has been improved in the last few decades, which helps to identify sphingolipid structures as well as to elucidate the biosynthetic pathway in different organisms. While glycosylceramides are present in all organisms studied, complex

Rajendra Prasad and Ashutosh Singh (eds.), *Analysis of Membrane Lipids*, Springer Protocols Handbooks, https://doi.org/10.1007/978-1-0716-0631-5_3,

sphingolipids, such as glycosyl inositol phosphoryl ceramide (GIPC), are found in fungi and plants, and sphingomyelin is exclusive to mammals [1, 3, 4].

Studies regarding sphingolipids biological functions have also been performed, highlighting the importance of this lipid class for cell membrane integrity, morphology, and growth, playing crucial roles in cellular processes, such as signal transduction, nutrient uptake by endocytosis, and regulation of cell proliferation. In addition, sphingolipids take part in defense processes in plants and mammals, such as the human immune system and the plant responses, and of virulence in pathogenic microorganisms, being essential for fungal infections in plants and mammalian cells [5–7].

This chapter aims to go over the main aspects of sphingolipid structures in different organisms and the main methods for lipid extraction, purification, and chemical analysis. In addition, the most relevant biological functions for mammals, plants, and fungi will be described.

2 Structural Aspects of Sphingolipids

The basis of sphingolipid structures is a sphingoid backbone composed of an eighteen-–carbon-alcohol chain that serves as a long-chain base (LCB) for the formation of the whole sphingolipids present in mammals, plants, and fungi. The variation in hydroxylation and saturation level results in different LCBs, such as sphingosine, phytosphingosine, and dihydrosphingosine [2]. In mammals, 4-sphingenine, with a double bond in carbon 4, is predominantly found as LCB, while 4,8-sphingadienine (two double bonds in carbons 4 and 8) is found in plants, and 9-methyl-4,8-sphingadienine (an extra methylation in carbon 9) is observed in fungi [3, 8] (Fig. 1).

From sphingoid bases, a variety of structures can be formed. For example, sphingosine-1-phosphate is produced through the phosphorylation of sphingosine, while its acylation results in ceramide structures, which is the base of many sphingolipids found in different organisms [2]. In this context, ceramides are composed of a fatty acid chain linked to the sphingoid backbone and the variation in this structure is based on the fatty acid length, hydroxylation, and saturation level. Mammalian fatty acid length is predominantly a C16 or C18, while in fungi C16–18 (in glycosylceramides) and C24–26 (in inositol phosphoryl ceramide) are observed [1]. In plants, the fatty acid length ranges from C14 to C26 [3].

A variety of molecules are synthesized from ceramide structure. Glycosylceramide is formed when a sugar unit is added to ceramide through a glycosidic bond [9]. Interestingly, in mammals and fungi glucose and galactose are found, while in plants glucose and

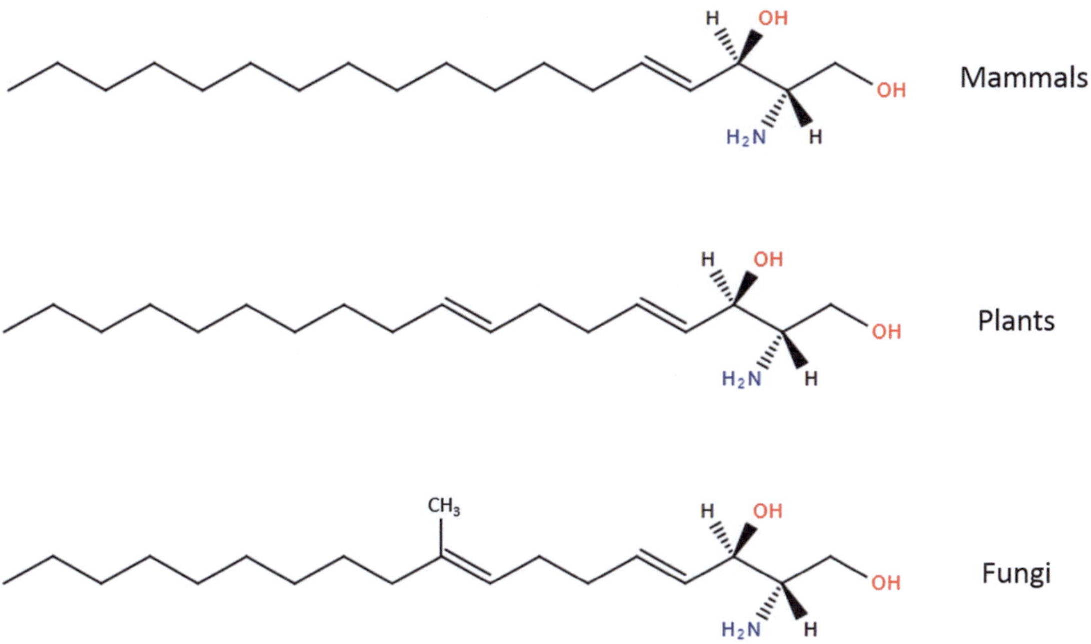

Fig. 1 Main long-chain bases found in mammals, plants, and fungi

mannose are observed [1]. In addition, plants also produce di-, tri-, and tetra-glycosylceramides [3].

Other sphingolipids are also present in these organisms. Fungi and plants synthesize glycosyl inositol phosphoryl ceramide (GIPC), in which an inositol phosphate is linked to the ceramide backbone [10]. Generally, it is composed of mannose as a sugar unit and very long fatty acid chains up to C26. Different forms of GIPCs are found, such as IPC (without sugar unit), MIPCs (mannose as a sugar unit), and MIP_2C (with two inositol groups) [11].

Gangliosides and sphingomyelin are typically found in mammalian cells and absent in plants and fungal cells. Gangliosides are acidic glycosphingolipids containing sialic acid residues in the sugar unit [12]. Depending on the amount of sialic acid residue (0, 1, 2, or 3), they are classified as asialo-, a-, b-, and c-series gangliosides, respectively [12]. These lipids are more abundant in the nervous system, localized on the outer face of the plasma membrane [13]. Sphingomyelin differs from glycosylceramide by the presence of a phosphorylcholine headgroup instead of the sugar unit linked to ceramide [14]. Acyl group variation is observed in sphingomyelin, depending on the tissue. For example, in the brain, a nervonic acid (24:1) predominates in the white matter, while an oleic acid (18:1) composes the grey matter [15, 16].

To sum up, sphingolipids are composed of an LCB, which is linked to a fatty acid to form ceramide resulting in the backbone for different molecules. From this, glycosylceramides, glycosyl inositol phosphoryl ceramides, gangliosides, and sphingomyelin are synthesized. In the next sections, we will discuss how glycosylceramides (the most studied sphingolipid in the literature) are extracted and purified from live samples, as well as the main biological aspects of these molecules in mammals, plants, and fungi. Figure 2 shows the main sphingolipid structures found.

3 Methods for Glycosphingolipid Structural Analysis

A variety of methodologies have been described in the literature to analyze sphingolipids of different types of cells. Basically, there are three main steps needed to obtain these molecules and identify their structures, as pointed below and illustrated by Barreto-Bergter and colleagues [9] (Fig. 3):

1. Extraction of total lipids from a certain amount of cells, resulting in the crude lipid extract.
2. Purification procedures to separate glycosphingolipids from the crude lipid extract.
3. Spectrometric and spectroscopic methods to identify glycosphingolipid structures.

The above steps vary significantly depending on the study and there are many procedures described in the literature for these reasons. In this section, different methods for glycosphingolipid analysis will be discussed.

3.1 Extraction of Total Lipids

This step aims to extract lipids from a cell sample in order to obtain the crude lipid extract, which contains different classes of lipids and from which sphingolipids are further purified. The sample amount necessary for lipid extraction depends on which organisms are studied and the lipid amount required for the study.

In general, organic solvents are used for lipid extraction, but there are plenty of mixtures described in the literature. Basically, there are two most common solutions used: one based on chloroform and methanol and another on ethanol, diethyl ether, and pyridine. Some examples of methods and references are listed in Table 1. Mandala and colleagues described a protocol composed of a mixture of ethanol:H_2O:diethylether:pyridine:NH_4OH (15:15:5:1:0.018 v/v) with incubation at 60 °C for 1 h [17]. Similarly, Smith and Lester used one volume of 95% ethanol, 1/3 volume of diethyl ether, and 1/15 volume of pyridine with incubation at 57 °C for 30 min with frequent agitation [10]. Methodologies based on chloroform and methanol were described by Toledo and colleagues [18], who used two solvents to extract lipids: a solvent A composed by isopropanol/hexane/water (55:20:25 v/v) and a solvent B composed by chloroform/methanol (2:1, v/v). Takakuwa et al. also performed lipid extraction using chloroform/

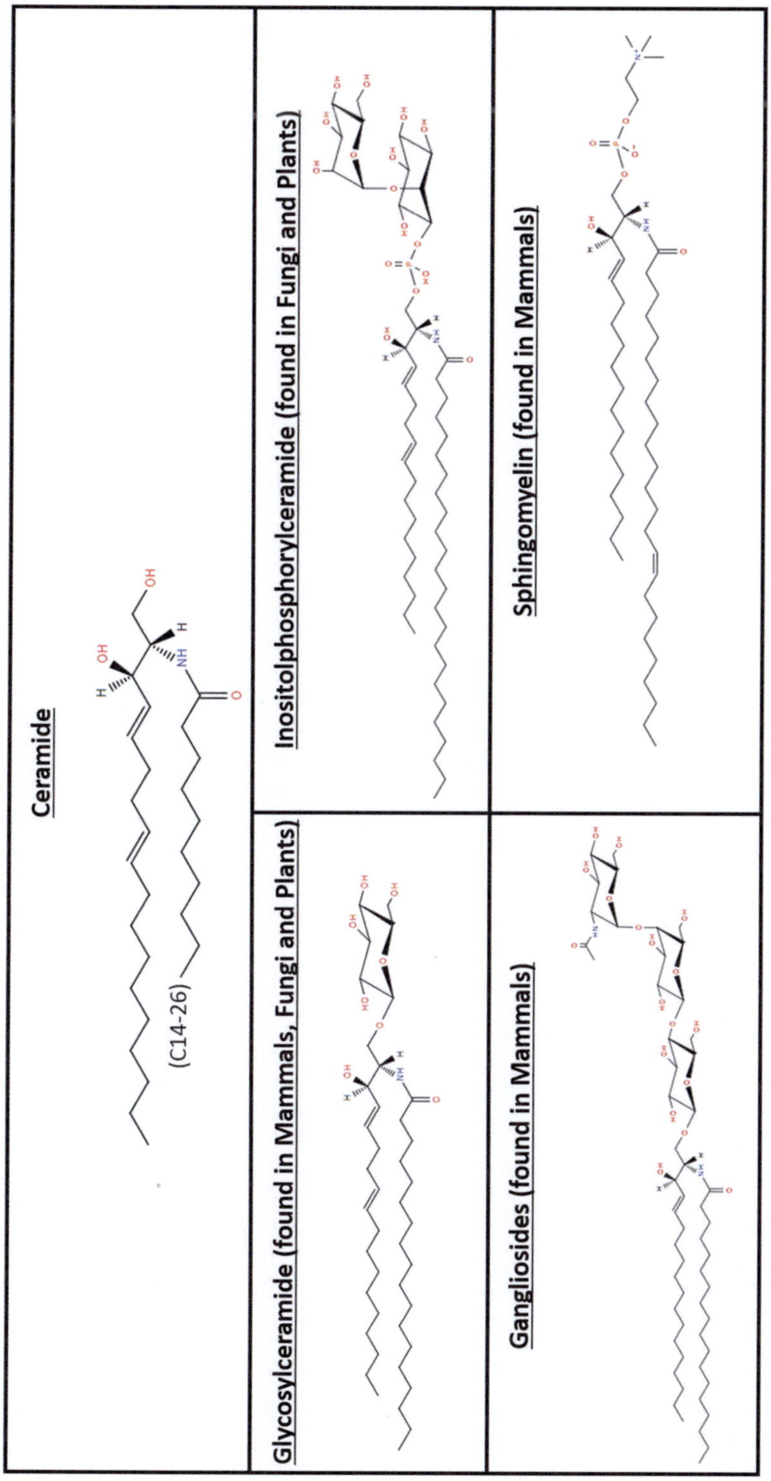

Fig. 2 Main sphingolipids found in mammals, plants, and fungi

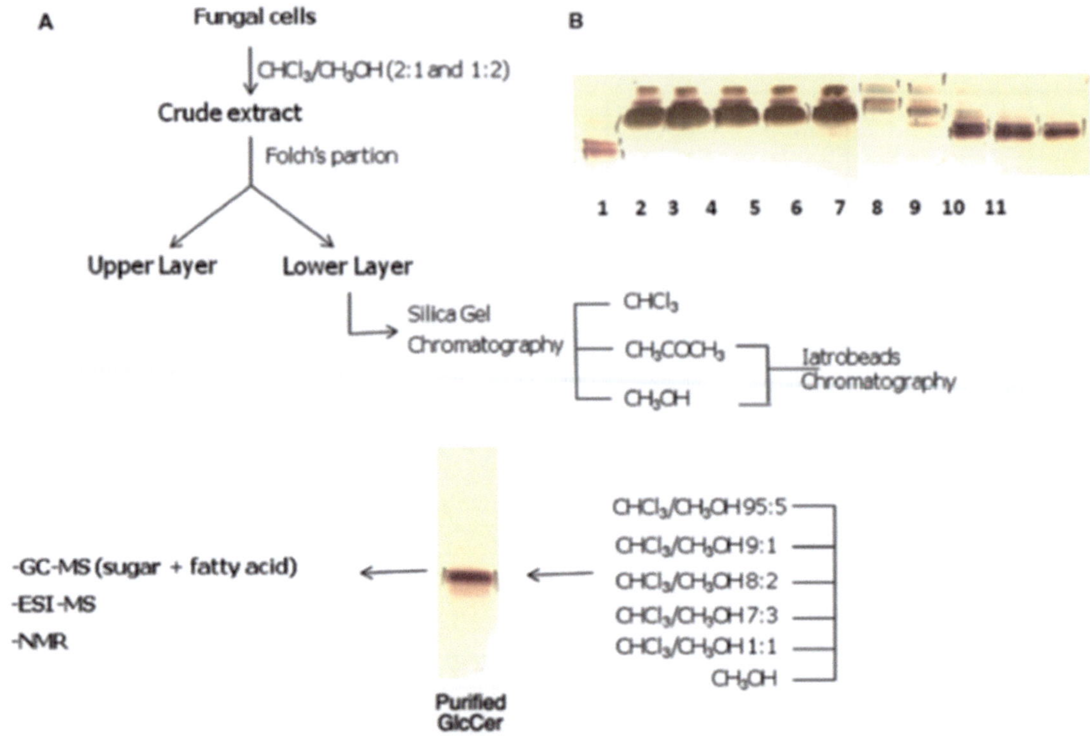

Fig. 3 Extraction, purification, and identification methods for sphingolipids, based specifically on glucosylceramide. (**a**) Purification steps to obtain glucosylceramides from crude lipid extract. (**b**) Thin-layer chromatography (TLC) of glycosphingolipids isolated by silica gel column chromatography: Lane1, GlcCer standard; Lanes 2–6: fractions from chloroform–methanol (95:5 v/v) elution; Lanes 7–8: fractions from chloroform–-methanol (90:10 v/v) elution; Lanes 9–11: fractions from chloroform–methanol (80:20 v/v) elution. Scheme found on Barreto-Bergter and colleagues [9]

Table 1
Different lipid extraction methods described in the literature

Extraction method	References
Ethanol:H_2O:diethylether:pyridine:NH_4OH (15:15:5:1:0.018; v/v) at 60 °C for 1 h	Mandala et al. [17]
One volume of 95% ethanol, 1/3 volume of diethyl ether, and 1/15 volume of pyridine	Smith and Lester [10]
Sol. A: isopropanol/hexane/water (55:20:25) Sol. B: chloroform/methanol (2:1, v/v)	Toledo et al. [18]
Chloroform:methanol (2:1, v/v) for 5 min by an ultrasonic disrupter	Takakuwa et al. [19]
Chloroform and methanol (2:1 v/v for 2 h, followed by 1:2 v/v) at room temperature	Barreto-Bergter et al. [9]

methanol (2:1, v/v), but with 5-min incubation in an ultrasonic disrupter [19]. The group of Barreto-Bergter and colleagues has been extracting crude lipids from fungal cells in the last few decades with chloroform and methanol (2:1 v/v for 2 h, followed by 1:2 v/v) at room temperature, obtaining a significant amount of lipid from different fungal species [9].

3.2 Purification Procedures

After the obtention of crude lipids, the samples are submitted to a partitioning step using a mixture of organic solvents aiming to separate neutral lipids in the lower phase (such as sphingolipids) and acidic lipids in the upper phase (such as sulfate lipids). Chloroform/methanol/water (30:60:8, v/v) has been used to generate a double phase mixture: chloroform in the lower phase and methanol/water in the upper phase [18, 20]. Bligh and Dyer described a method using methanol:chloroform (2:1;v/v) solvent, with the further addition of 1/3 volume of chloroform and H_2O in glass tubes [21]. Barreto-Bergter and colleagues have been using a protocol based on Folch partitioning [22], in which the lipids are dissolved with chloroform, methanol, and 0.75% KCl overnight in a glass decanter container [9].

Chromatography columns have been used to purify sphingolipids after the partitioning step. In general, a variety of elution schemes have been found in the literature. Takahashi and colleagues, for example, used Iatrobeads 8010 column eluted with isopropanol/hexane/water (55:43:2 to 55:20:25) [20]. On the other hand, Barreto-Bergter and Toledo based their chromatographic methods on a silica gel 60 column eluted with chloroform, acetone, and methanol or with a gradient of chloroform:methanol from 9:1 to 1:1 (v/v) [9, 18].

In order to check the purity of sphingolipids during the steps above, thin-layer chromatography is performed, in which lipid samples are spotted on a silica gel layer located on a glass plate. Different elution solvents to carry lipids along silica gel layer are used, such as chloroform:methanol (95:12, v/v) [19], chloroform/methanol/water (60:40:9 v/v/v) [18], or chloroform/methanol/2 M NH_4OH (40:10:1 v/v/v) [9].

Table 2 summarizes the main techniques used for sphingolipid purification.

3.3 Identification of Purified Sphingolipids

After obtaining purified sphingolipids by the methods described above, spectrometric and spectroscopic techniques have been used to identify their structures. Mass spectrometry is by far the best technique to characterize sphingolipid structures. In this context, lithium hydroxide (LiOH) or lithium chloride (LiCl) is added in the mixture before the injection of lipid samples in the spectrometry [23]. A variety of ionization options can be used in sphingolipid analysis, such as ESI, MALDI, and FAB, usually coupled with mass analyzers such as MS/MS and TOF [24–28]. Among them,

Table 2
Several sphingolipid purification methods described in the literature

Purification method	References
Partitioning with chloroform/methanol/water (30:60:8, v/v)	Toledo et al. [18] Takahashi et al. [20]
Partitioning with methanol:chloroform (2:1;v/v) solvent + 1/3 volume of chloroform and H_2O in glass tubes	Bligh and Dyer [21]
Partitioning with chloroform, methanol, and 0.75% KCl overnight in a glass decanter container	Folch [9] Barreto-Bergter et al. [22]
Column chromatography: Iatrobeads 8010 column eluted with isopropanol/hexane/water (55:43:2 to 55:20:25 v/v/v)	Takahashi et al. [20]
Column chromatography: Silica gel 60 column eluted with chloroform, acetone and methanol, or with a gradient of chloroform:methanol from 9:1 to 1:1 (v/v)	Toledo et al. [9] Barreto-Bergter et al. 2011 [18]

ESI-MS/MS is the most used approach to study sphingolipid structures. Lipid samples are dissolved in organic solvents and injected into the ESI source via a narrow capillary needle. ESI is a soft ionization method that generates intact positively or negatively charged ions, formed by using high voltage (positive or negative), vacuum, and N2, producing a gaseous phase with charged molecular ions that are detected by the analyzer [29, 30].

Mass spectrometry is usually coupled to high-performance liquid chromatography (HPLC) or gas chromatography (GC), forming HPLC-ESI-MS/MS or GCMS, the latter being one of the most used techniques for sphingolipids. These approaches allow molecular ion separation based on m/z (mass-to-charge), which is commonly found in lipid studies. Fragmentation patterns of sphingolipids are present in the literature, which makes now easier for us to identify the molecular ions found in our studies and, thus, to structurally characterize new sphingolipid molecules [30, 31].

4 Biological Aspects of Mammalian Sphingolipids

Basically, there are two typically abundant mammalian sphingolipids: gangliosides and sphingomyelin. For this reason, this section will focus on the biological roles of these two groups of molecules.

Sphingolipids present a variety of functions in mammals, and one of the best-studied properties is their role in cellular signaling. Ceramides and sphingomyelin comprise approximately 50% of lipid microdomains in the plasma membrane, called lipid rafts, which are centers for recruitment and localization of receptors and signaling molecules [32]. The long and saturated carbon chains typical of sphingolipid structures allow tight packing of the membrane and give more stability to anchored proteins in this region [33]. Structural modifications in receptors and signaling proteins lead to their anchoring to lipid rafts, such as glycosylinositolphosphate (GPI) addition, which allows protein insertion on the outer membrane leaflet, and palmitoylation and farnesylation, which allow the protein interaction with the inner face of the plasma membrane [34–37]. A variety of proteins are associated with microdomains, where they promote their biological effects, such as IgG receptors, Src family tyrosine kinase, growth factor receptors, G protein-coupled receptors, extracellular regulated kinases (ERK1 and 2), endothelial nitric oxide synthase, integrins, T-cell receptors, and TNF receptor [34, 36–41].

Caveolin is a protein associated with membrane microdomain formed by sphingomyelin and cholesterol. Mice lacking caveolin-1 are more resistant to diet-induced obesity, suggesting that caveolae are important membrane domains for endocytosis and nutrient uptake, as well as that sphingomyelin participates in the regulation of body metabolism [42–44]. In addition, these lipids are related to iron uptake, since the plasma membrane possesses receptors that bind to transferrin, which is crucial for cell proliferation [45, 46]. Studies using lymphoma cells observed that disrupting sphingomyelin blocked cell growth. On the other hand, transfecting these cells with sphingomyelin synthase (SMS1 and SMS2) restored transferrin endocytosis and cell proliferation [47].

Sphingomyelin-containing microdomains are also associated with anchoring voltage-activated potassium (kV) channels, which are essential for nerve impulse. The use of sphingomyelinase C and D, disrupting sphingomyelin content, led to an inhibition of certain types of kV channels, suggesting that these lipids play a role in regulating some types of kV channels, helping information transference between nervous cells [48].

Gangliosides are associated with a variety of cellular processes, such as cell–cell interaction, adhesion and signaling, as well as calcium homeostasis [12, 49]. The best tool to study ganglioside importance for mammals is the development of mutant mice by knocking out genes of ganglioside synthases, such as sialyltransferases (STs) and *N*-acetylgalactosaminyltransferase (GalNAcT). Since gangliosides are strictly associated with nervous tissues, many phenotype alterations are related to nervous system disorders [12]. Mice lacking ST-I presented a dysfunction in cochlear maturation, leading to deafness [50]. In addition, they exhibit attention-deficit hyperactivity disorder, suggesting the role of gangliosides in neuropsychological balance [51]. ST-II disruption led to the loss of nerve regenerative ability [52]. GalNAcT gene deletion led to abnormal neuronal transmission, since a reduction of nerve conduction velocity was observed [53]. When double mutants were developed, weight and memory loss were observed, as well as a high lethality level [54–56].

Many diseases involve gangliosides in its development. Guillain-Barré syndrome, for instance, is an autoimmune process in which cell surface gangliosides are recognized by the immune system [57]. In Alzheimer's disease, the ganglioside Chol-1α causes the amyloid-β peptide aggregation and, thus, the degenerative symptoms of the pathology [58, 59]. Moreover, gangliosides are useful as biomarkers for stem cells, helping to identify and isolate them, and for brain cancer cells, which express c-series gangliosides in their surface [60–63].

5 Biological Aspects of Plant Sphingolipids

Sphingolipids along with their phosphorylated derivatives are bioactive components of plant cells. They are structural elements in the lipid bilayer and contribute to the dynamic nature of the plasma membrane. Although sphingolipids have a better-defined role in animals, they are important for many essential processes in plants, such as pollen development, signal transduction, programmed cell death, response to biotic and abiotic stress, in the response of plants to hypoxia, and to the attack of pathogens [5]. Sphingolipids constitute approximately 40% of the total lipids in the plasma membrane of plants [64], mostly localized in the outer leaflet. Recent data show that plant sphingolipids are composed of four main classes: ceramides (CER), glycosylceramides (GlyCer), glycosyl inositol phosphoryl ceramides (GIPC), and free chain bases (LCBs), representing about 2%, 34%, 64%, and 0.5% of total sphingolipids, respectively, in *Arabidopsis thaliana* [65, 66]. In addition, they are also abundant lipid components of other endomembranes in the plant, including endoplasmic reticulum, Golgi apparatus, and tonoplasts [67].

Studies have shown that the lipid composition of the plasma membrane is altered when plants are exposed to drought, heavy metals, cooling, and freezing tolerance. The main cause of freezing injury in plants is membrane destabilization, resulting in osmotic stress and dehydration. The process of acclimatization to cold conditions in many plant species is accompanied by changes in the lipid composition of the plasma membrane and a decrease in the proportion of glucosylceramides [5, 68].

In the evolutionary process, the male gametophyte of the angiosperms was reduced to the pollen grain, being a vegetative cell with two spermatozoa. In order to deliver the sperm from the stigma through the stylet into the ovule, the pollen tube is synthesized and undergoes an elongation process in some species, requiring large amounts of fatty acids and membrane lipid synthesis. A special lipid composition in pollen has been acquired over the years to make the pollen tube growth a more efficient process. Pollen tubes produce extra plastid galactolipids and store triacylglycerols in lipid droplets, probably needed as glycerolipid precursors or for acyl editing. They also have a special moiety of sterol and sphingolipids that together can form microdomains in the membranes [69]. In *Arabidopsis*, a glycosylceramide synthase (GCS) is encoded by a single essential gene, whereas GIPCs are synthesized using three functional components of IPC and several glycosyltransferase or glucuronyltransferases [5, 70, 71]. Reports show that functional GCS is not able to develop beyond the seedling stage. However, heterozygous plants carrying a null GCS allele had transmission of defective pollen [5, 72].

The first synthase of the identified IPC was ERH1, found in screening for mutants of increased hypersensitivity response during pathogen infection [73]. The ERH1 mutant showed no obvious change in GIPC levels. However, it accumulated substrate ceramides, and this was the likely cause of increased cell death responses in the mutant. The additional glycan ornamentation of GIPC was elucidated with the identification of IPUT1 as inositol phosphorylceramide glucuronosyltransferase 1, which transfers a glucuronic acid (GlcA) to GIPC [5, 71]. The lack of a single glycan such as GlcA in GIPC has an impact because it prevented the transmission of pollen [5].

Membrane microdomains (rafts) rich in sphingolipids and sterols were found in cell membranes (especially the plasma membrane) of eukaryotic cells [33, 74]. Results from Peskan and colleagues [75] show that microdomains isolated from the plasma membrane of tobacco leaves contain a significant fraction of β-subunit coupled to the membrane of G-protein heterotrimeric and six polypeptides enriched identified as GPI-anchored proteins [75]. Moreover, some proteins with signaling functions, such as protein G, calcium signaling proteins, and receptor kinases (FLS2, for instance), have already been described as members of plant lipid rafts [76, 77]. The sphingolipid abundance in the plasma membrane is linked with its role in the formation of membrane microdomains [5]. These microdomains are not larger than 35 nm, are highly resistant to detergent, and showed strong enrichment in highly glycosylated GIPC containing VLCFAs [5].

In addition, research also suggests the role of sphingolipids in response to biotic stress against herbivores. Begum and colleagues [78] showed that the OsLCB2a1 gene encoding a subunit of serine palmitoyltransferase (SPT) protects plants against the attack and infection of the herbivore brown planthopper (BPH). Analysis of gene expression in rice seedlings shows levels of transcription of the OsLCB2a1 gene at 4 h, but decreased at 8–24 h after the BPH attack [79].

6 Biological Aspects of Fungal Sphingolipids

Ergosterol and sphingolipids are the main lipid components of microdomains in fungal membranes. These regions play crucial roles in cellular signaling, fungal growth, and virulence [4, 8, 80]. Regarding cellular signaling processes, sphingolipid synthesis interferes in some protein kinases. In *Cryptococcus neoformans*, the enzyme Ipc1 performs IPC synthesis, transferring phosphorylinositol moiety to phytoceramide and releasing diacylglycerol (DAG). DAG binds to and increases the activity of cryptococcal protein kinase 1 (Pkc1), which is required for cell wall integrity and the function of some enzymes, such as laccase, an important component of melanin synthesis. Thus, it is described that Ipc1 downregulation impairs melanin production in *C. neoformans* [81].

Sphingolipids are also required for hyphal growth and cell polarization, which is mediated by the cytoskeleton and the accumulation of lipid raft components at the growing tip of hyphal cells. *Candida albicans* treatment with myriocin, an inhibitor of serine palmitoyltransferase, leads to the decrease in lipid domain accumulation, resulting in less recruitment of heat-shock proteins and molecules required for sterol metabolism, energy production, and polysaccharide synthesis [82]. In *Aspergillus nidulans* and *Fusarium graminearum* lacking enzymes of sphingolipid synthesis, such as Bar1p, it is observed an alteration in hyphal tip organization due to a reduction in sterol accumulation and the presence of irregular spitzenkörper formation [80].

In addition to cell polarization and hyphal growth, sphingolipids are crucial for extracellular vesicle formation and, consequently, for polysaccharide and protein sorting [83]. It influences the secretion of important constituents for the formation of cell surface structures in fungi, such as cell wall polysaccharides and the capsule components in *C. neoformans* [84].

The use of mutants lacking enzymes of the sphingolipid biosynthetic pathway is a useful approach to understand the biological roles of these molecules for fungal growth and virulence. Knockout strains have already been developed for *C. neoformans*, *A. nidulans*, *F. graminearum*, *P. digitatum*, and *C. albicans* [85–89]. All these works demonstrated that deleting genes involved in sphingolipid synthesis, such as *GCS1* (glucosylceramide synthase), *MTS1* (C9-methyltransferase), and *SLD* (Δ^8 desaturase), leads to decreased virulence in an animal infection model, as well as defects in fungal growth and hyphal elongation. Interestingly, intact glucosylceramide was crucial for *C. neoformans* growth in alkali/neutral conditions found in host extracellular environments, such as the alveolar space and bloodstream, but it was not essential for growing in acidic conditions found in intracellular environments, such as the macrophage phagolysosomes, suggesting that sphingolipids play important roles in alkali tolerance and the fungal adaptation to host sites [85]. Also, changes in glucosylceramide structures result in a higher susceptibility of *C. neoformans* to membrane stressors [90]. In *C. albicans*, the deletion of the *SLD* gene led to increased susceptibility to SDS (sodium dodecyl sulfate) and fluconazole, indicating that disruption of sphingolipid synthesis weakens plasma membrane [89]. In plant infection models, *F. graminearum* and *P. digitatum* knockout strains also displayed less virulence, suggesting that sphingolipids play crucial roles in virulence not only in animal infections but also in phytopathogenic fungi [86, 87].

The effect of synthetic sphingolipid inhibitors in fungal biological processes has already been demonstrated in the literature for a variety of fungi. Myriocin and sphingofungin, inhibitors of serine palmitoyltransferase (SPT), the first step of sphigolipid biosynthetic pathway, have been shown to block biofilm formation by *Candida* and *Aspergillus* species [91, 92]. Ceramide synthase inhibitors, such as fumonisin B1 and australifungin, presented antifungal activities in *Cryptococcus*, *Candida*, and *Aspergillus* species [17]. IPC synthases, especially the enzyme Ipc1, are inhibited by aureobasidin A, khakrefungin, and galbonolide. In *C. albicans*, *C. neoformans*, and *Saccharomyces cerevisiae*, these IPC inhibitors present antifungal activity by leading to ceramide accumulation to toxic levels, impairing biofilm formation and germ tube elongation [93, 94]. Glucosylceramide synthases (GCS) take part in the last step of glucosylceramide synthesis and are blocked by D-*threo*-1-phenyl-2-decanoylamino-3-morpholino-1-propanol (D-*threo*-PDPM). In *A. nidulans* and *A. fumigatus*, this inhibitor was able to reduce hyphal germination and colony growth [95].

Molecules targeting directly the sphingolipids on the fungal cell surface have also been described. A variety of natural products isolated from plants and insects, such as the defensins MsDefl and RsAFP2, can bind sphingolipids on cell surface and be active against *C. albicans* and *Pichia pastoris*, causing alterations in cell shape and growth, as well as blocking yeast–hyphae transition, an essential step for *Candida* pathogenesis [96, 97]. Interestingly, *Candida glabrata*, which does not produce glucosylceramide, and mutants lacking this molecule are not susceptible to these defensins, suggesting that sphingolipids are the target of these natural products [98]. In addition, MsDefl and RsAFP2 do not recognize human sphingolipids, indicating that structural differences between mammalian and fungal lipids are crucial for selective binding of defensins [96]. Antibodies against glucosylceramides have been used as tools to study the roles of sphingolipids. Rodrigues and colleagues obtained sera from patients presenting cryptococcosis and observed the presence of antibodies against glucosylceramides which could inhibit fungal bud formation and protected mice in an in vivo infection assay, indicating that this molecule possesses an immunogenic property [99, 100]. In

Lomentospora prolificans, purified glucosylceramide recruited different immune cells and induced the production of pro-inflammatory cytokines and the oxidative burst by macrophages, indicating that glucosylceramides are potent immune response activators [101]. Monoclonal antibodies have also been used in a variety of in vitro studies, causing a reduction in fungal growth and germination, as well as enhancing macrophage fungicidal activity, as observed for *Fonsecaea pedrosoi*, *Colletotrichum gloeosporioides*, *Pseudallescheria boydii*, and *Scedosporium apiospermum* [102–105].

In the context of all the data showed above, sphingolipids have been considering in the last few decades a potent new target for the development of a new class of antifungal drugs, not only due to the structural differences between mammals and fungi but also due to the biological properties presented here [4]. Recently, a group of chemically synthetic compounds, such as *N'*-(3- bromo-2-hydroxybenzylidene)-2-methylbenzohydrazide (BHBM) and 3-bromo-*N'*-(3-bromo-4-hydroxybenzylidene) benzohydrazide (D0), was discovered to display in vitro and in vivo antifungal activity against *C. neoformans*, *Candida* species, *Aspergillus fumigatus*, and *Histoplasma capsulatum* [106]. Chemical modifications in these acylhydrazone molecules have been performed and new derivatives from BHBM are shown to be active against *C. neoformans* and to present low toxicity in mammalian cells [107]. These data support the importance of studying sphingolipids in fungi, especially aiming to clarify the roles for fungal biology and to use these molecules as a new therapeutic option in the future.

In the context of fungi–plant interaction, sphingolipids also play important roles. Fungal glucosylceramides have already been investigated due to their ability to confer resistance to plant pathogens and induce immunity with different defense mechanisms. The fungal glucosylceramides act as an elicitor to stimulate the biosynthesis of phytoalexins and PR proteins when sprayed on rice leaves [108]. Researchers have shown that *Magnaporthe oryzae* cerebrosides A and C induce hypersensitive response and defense in rice plants [109, 110]. Spraying rice leaves with cerebroside A induced the accumulation of phytoalexins (antimicrobial substances) and increased resistance to a subsequent one by compatible pathogens. Koga [109] has shown that cerebroside degradation products do not have an eliciting activity. However, Umemura et al.'s [110] studies have generated ceramides prepared from cerebrosides through the removal of glucose which present the ability to elicit defense activity. In addition, the methyl group at C-9 and the double bond at the sphingoid base of cerebrosides A and C are the key elements that determine the eliciting activity of these compounds [109]. Umemura and colleagues also detected cerebrosides A, B, and C in several soil-borne phytopathogens, such as *Fusarium*, *Pythium*, and *Botrytis* species, and found that sphingolipids acting as defense elicitors are not important in propagating plants of different species [111]. However, the mechanism by which these cerebrosides derived from pathogens induce defense response is currently unknown. There are two possible mechanisms: (1) these cerebrosides or their ceramide metabolites may interfere with the sphingolipid metabolism of host plants, causing SPI-PCD (sphingolipid-perturbation induced PCD) only as exogenous free ceramides or LCBs (long-chain base), or (2) these pathogen-specific lipids are recognized by host plants as PAMPs (pathogen-associated molecular pattern) by the immune receptor protein in the plasma membrane, thereby inducing immune response [79].

7 Conclusions

As described above, sphingolipids are a diverse class of lipids and the study of its structural variations could reveal important correlations between structure and biological functions, since it is observed among the various sphingoid structures presented in mammals, plants, and fungi. In addition, studying sphingolipids is essential to understand many biological processes, such as membrane integrity, cell proliferation, and nutrient uptake. These acknowledge could serve as useful tools for clarifying the development of some diseases in humans, such as neurodegenerative syndromes and fungal infections, as well as the development of new therapeutic approaches.

References

1. Marques JT, Marinho HS, De Almeida RFM (2018) Sphingolipid hydroxylation in mammals, yeast and plants – an integrated view. Prog Lipid Res 71:18–42
2. Gault CR, Obeid LM, Hannun YA (2010) An overview of sphingolipid metabolism: from synthesis to breakdown. Adv Exp Med Biol 688:1–23
3. Sperling P, Heinz E (2003) Plant sphingolipids: structural diversity, biosynthesis, first genes and functions. Biochim Biophys Acta 1632(1–3):1–15
4. Rollin-Pinheiro R, Singh A, Barreto-Bergter-E, Del Poeta M (2016) Sphingolipids as targets for treatment of fungal infections. Future Med Chem 8(12):1469–1484
5. Michaelson LV, Napier JA, Molino D, Faure JD (2016) Plant sphingolipids: their importance in cellular organization and adaption. Biochim Biophys Acta 1861(9 Pt B):1329–1335
6. Slotte JP (2013) Biological functions of sphingomyelins. Prog Lipid Res 52 (4):424–437
7. Heung LJ, Luberto C, Del Poeta M (2006) Role of sphingolipids in microbial pathogenesis. Infect Immun 74(1):28–39
8. Barreto-Bergter E, Pinto MR, Rodrigues ML (2004) Structure and biological functions of fungal cerebrosides. An Acad Bras Cienc 76 (1):67–84
9. Barreto-Bergter E, Sassaki GL, De Souza LM (2011) Structural analysis of fungal cerebrosides. Front Microbiol 2:239
10. Smith SW, Lester RL (1974) Inositol phosphorylceramide, a novel substance and the chief member of a major group of yeast sphingolipids containing a single inositol phosphate. J Biol Chem 249(11):3395–3405
11. Leber A, Fischer P, Schneiter R, Kohlwein SD, Daum G (1997) The yeast mic2 mutant is defective in the formation of mannosyl-

diinositolphosphorylceramide. FEBS Lett 411(2–3):211–214

12. Yu RK, Tsai YT, Ariga T, Yanagisawa M (2011) Structures, biosynthesis, and functions of gangliosides--an overview. J Oleo Sci 60(10):537–544
13. Yu RK, Nakatani Y, Yanagisawa M (2009) The role of glycosphingolipid metabolism in the developing brain. J Lipid Res 50:S440–S445
14. Airola MV, Hannun YA (2013) Sphingolipid metabolism and neutral sphingomyelinases. Handb Exp Pharmacol 215:57–76. https://doi.org/10.1007/978-3-7091-1368-4_3
15. O'brien JS, Sampson EL (1965) Lipid composition of the normal human brain: gray matter, white matter, and myelin. J Lipid Res 6(4):537–544
16. Ohanian J, Ohanian V (2001) Sphingolipids in mammalian cell signalling. Cell Mol Life Sci 58(14):2053–2068
17. Mandala SM, Thornton RA, Frommer BR, Curotto JE, Rozdilsky W, Kurtz MB, Giacobbe RA, Bills GF, Cabello MA, Martín I, Pelaez F, Harris GH (1995) The discovery of australifungin, a novel inhibitor of sphinganine N-acyltransferase from *Sporormiella australis*. Producing organism, fermentation, isolation, and biological activity. J Antibiot (Tokyo) 48(5):349–356
18. Toledo MS, Levery SB, Suzuki E, Straus AH, Takahashi HK (2001) Characterization of cerebrosides from the thermally dimorphic mycopathogen *Histoplasma capsulatum*: expression of 2-hydroxy fatty N-acyl (E)-Delta(3)-unsaturation correlates with the yeast-mycelium phase transition. Glycobiology 11(2):113–124
19. Takakuwa N, Kinoshita M, Oda Y, Ohnishi M (2002) Existence of cerebroside in *Saccharomyces kluyveri* and its related species. FEMS Yeast Res 2(4):533–538
20. Takahashi HK, Levery SB, Toledo MS, Suzuki E, Salyan ME, Hakomori S, Straus AH (1996) Isolation and possible composition of glucosylceramides from *Paracoccidioides brasiliensis*. Braz J Med Biol Res 29 (11):1441–1444
21. Bligh EG, Dyer WJ (1959) A rapid method of total lipid extraction and purification. Can J Biochem Physiol 37(8):911–917
22. Folch J, Lees M, Sloane Stanley GH (1957) A simple method for the isolation and purification of total lipides from animal tissues. J Biol Chem 226(1):497–509
23. Sullards MC, Liu Y, Chen Y, Merrill AH Jr (2011) Analysis of mammalian sphingolipids by liquid chromatography tandem mass spectrometry (LC-MS/MS) and tissue imaging mass spectrometry (TIMS). Biochim Biophys Acta 1811(11):838–853
24. Han X, Gross RW (1994) Electrospray ionization mass spectroscopic analysis of human erythrocyte plasma membrane phospholipids. Proc Natl Acad Sci U S A 91 (22):10635–10639
25. Jones EE, Dworski S, Canals D, Casas J, Fabrias G, Schoenling D, Levade T, Denlinger C, Hannun YA, Medin JA, Drake RR (2014) On-tissue localization of ceramides and other sphingolipids by MALDI mass spectrometry imaging. Anal Chem 86 (16):8303–8311
26. Ahn YM, Lee WW, Jung JH, Lee SG, Hong J (2009) Structural determination of glucosylceramides isolated from marine sponge by fast atom bombardment collision-induced dissociation linked scan at constant B/E. J Mass Spectrom 44(12):1698–1708
27. Shaner RL, Allegood JC, Park H, Wang E, Kelly S, Haynes CA, Sullards MC, Merrill AH Jr (2009) Quantitative analysis of sphingolipids for lipidomics using triple quadrupole and quadrupole linear ion trap mass spectrometers. J Lipid Res 50(8):1692–1707
28. Jia Z, Li S, Cong P, Wang Y, Sugawara T, Xue C, Xu J (2015) High throughput analysis of cerebrosides from the sea cucumber *Pearsonothria graeffei* by liquid chromatography-quadrupole-time-of-flight mass spectrometry. J Oleo Sci 64(1):51–60
29. Brugger B (2014) Lipidomics: analysis of the lipid composition of cells and subcellular organelles by electrospray ionization mass spectrometry. Annu Rev Biochem 83:79–98
30. Singh A, Del Poeta M (2016) Sphingolipidomics: an important mechanistic tool for studying fungal pathogens. Front Microbiol 7:501
31. Guan XL, Wenk MR (2006) Mass spectrometry-based profiling of phospholipids and sphingolipids in extracts from *Saccharomyces cerevisiae*. Yeast 23(6):465–477
32. Koval M, Pagano RE (1991) Intracellular transport and metabolism of sphingomyelin. Biochim Biophys Acta 1082(2):113–125
33. Simons K, Ikonen E (1997) Functional rafts in cell membranes. Nature 387 (6633):569–572
34. Janes PW, Ley SC, Magee AI, Kabouridis PS (2000) The role of lipid rafts in T cell antigen receptor (TCR) signalling. Semin Immunol 12(1):23–34
35. Anderson RG (1998) The caveolae membrane system. Annu Rev Biochem 67:199–225

36. Okamoto T, Schlegel A, Scherer PE, Lisanti MP (1998) Caveolins, a family of scaffolding proteins for organizing "preassembled signaling complexes" at the plasma membrane. J Biol Chem 273(10):5419–5422
37. Brown DA, London E (1998) Functions of lipid rafts in biological membranes. Annu Rev Cell Dev Biol 14:111–136
38. Igarashi J, Michel T (2000) Agonist-modulated targeting of the EDG-1 receptor to plasmalemmal caveolae. eNOS activation by sphingosine 1-phosphate and the role of caveolin-1 in sphingolipid signal transduction. J Biol Chem 275(41):32363–32370
39. Natoli G, Costanzo A, Guido F, Moretti F, Levrero M (1998) Apoptotic, non-apoptotic, and anti-apoptotic pathways of tumor necrosis factor signalling. Biochem Pharmacol 56 (8):915–920
40. Rosenman SJ, Ganji AA, Tedder TF, Gallatin WM (1993) Syn-capping of human T lymphocyte adhesion/activation molecules and their redistribution during interaction with endothelial cells. J Leukoc Biol 53(1):1–10
41. Bourguignon LY, Jy W, Majercik MH, Bourguignon GJ (1988) Lymphocyte activation and capping of hormone receptors. J Cell Biochem 37(2):131–150
42. Razani B, Wang XB, Engelman JA, Battista M, Lagaud G, Zhang XL, Kneitz B, Hou H Jr, Christ GJ, Edelmann W, Lisanti MP (2002) Caveolin-2-deficient mice show evidence of severe pulmonary dysfunction without disruption of caveolae. Mol Cell Biol 22(7):2329–2344
43. Ortegren U, Karlsson M, Blazic N, Blomqvist M, Nystrom FH, Gustavsson J, Fredman P, Strålfors P (2004) Lipids and glycosphingolipids in caveolae and surrounding plasma membrane of primary rat adipocytes. Eur J Biochem 271(10):2028–2036
44. Ortegren U, Aboulaich N, Ost A, Stralfors P (2007) A new role for caveolae as metabolic platforms. Trends Endocrinol Metab 18 (9):344–349
45. Gomme PT, Mccann KB, Bertolini J (2005) Transferrin: structure, function and potential therapeutic actions. Drug Discov Today 10 (4):267–273
46. Gatter KC, Brown G, Trowbridge IS, Woolston RE, Mason DY (1983) Transferrin receptors in human tissues: their distribution and possible clinical relevance. J Clin Pathol 36 (5):539–545
47. Shakor AB, Taniguchi M, Kitatani K, Hashimoto M, Asano S, Hayashi A, Nomura K, Bielawski J, Bielawska A, Watanabe K, Kobayashi T, Igarashi Y, Umehara H, Takeya H, Okazaki T (2011) Sphingomyelin synthase 1-generated sphingomyelin plays an important role in transferrin trafficking and cell proliferation. J Biol Chem 286(41):36053–36062
48. Milescu M, Bosmans F, Lee S, Alabi AA, Kim JI, Swartz KJ (2009) Interactions between lipids and voltage sensor paddles detected with tarantula toxins. Nat Struct Mol Biol 16 (10):1080–1085
49. Ledeen RW, Wu G (2008) Nuclear sphingolipids: metabolism and signaling. J Lipid Res 49(6):1176–1186
50. Yoshikawa M, Go S, Takasaki K, Kakazu Y, Ohashi M, Nagafuku M, Kabayama K, Sekimoto J, Suzuki S, Takaiwa K, Kimitsuki T, Matsumoto N, Komune S, Kamei D, Saito M, Fujiwara M, Iwasaki K, Inokuchi J (2009) Mice lacking ganglioside GM3 synthase exhibit complete hearing loss due to selective degeneration of the organ of Corti. Proc Natl Acad Sci USA 106 (23):9483–9488
51. Niimi K, Nishioka C, Miyamoto T, Takahashi E, Miyoshi I, Itakura C, Yamashita T (2011) Impairment of neuropsychological behaviors in ganglioside GM3-knockout mice. Biochem Biophys Res Commun 406 (4):524–528
52. Okada M, Itoh Mi M, Haraguchi M, Okajima T, Inoue M, Oishi H, Matsuda Y, Iwamoto T, Kawano T, Fukumoto S, Miyazaki H, Furukawa K, Aizawa S, Furukawa K (2002) B-series ganglioside deficiency exhibits no definite changes in the neurogenesis and the sensitivity to Fas-mediated apoptosis but impairs regeneration of the lesioned hypoglossal nerve. J Biol Chem 277 (3):1633–1636
53. Takamiya K, Yamamoto A, Furukawa K, Yamashiro S, Shin M, Okada M, Fukumoto S, Haraguchi M, Takeda N, Fujimura K, Sakae M, Kishikawa M, Shiku H, Furukawa K, Aizawa S (1996) Mice with disrupted GM2/GD2 synthase gene lack complex gangliosides but exhibit only subtle defects in their nervous system. Proc Natl Acad Sci U S A 93 (20):10662–10667
54. Tajima O, Egashira N, Ohmi Y, Fukue Y, Mishima K, Iwasaki K, Fujiwara M, Inokuchi J, Sugiura Y, Furukawa K, Furukawa K (2009) Reduced motor and sensory functions and emotional response in GM3-only mice: emergence from early stage of life and exacerbation with aging. Behav Brain Res 198 (1):74–82

55. Tajima O, Egashira N, Ohmi Y, Fukue Y, Mishima K, Iwasaki K, Fujiwara M, Sugiura Y, Furukawa K, Furukawa K (2010) Dysfunction of muscarinic acetylcholine receptors as a substantial basis for progressive neurological deterioration in GM3-only mice. Behav Brain Res 206(1):101–108
56. Yamashita T, Wu YP, Sandhoff R, Werth N, Mizukami H, Ellis JM, Dupree JL, Geyer R, Sandhoff K, Proia RL (2005) Interruption of ganglioside synthesis produces central nervous system degeneration and altered axon-glial interactions. Proc Natl Acad Sci U S A 102(8):2725–2730
57. Kaida K, Ariga T, Yu RK (2009) Antiganglioside antibodies and their pathophysiological effects on Guillain-Barre syndrome and related disorders--a review. Glycobiology 19 (7):676–692
58. Bernardo A, Harrison FE, McCord M, Zhao J, Bruchey A, Davies SS, Jackson Roberts L 2nd, Mathews PM, Matsuoka Y, Ariga T, Yu RK, Thompson R, McDonald MP (2009) Elimination of GD3 synthase improves memory and reduces amyloid-beta plaque load in transgenic mice. Neurobiol Aging 30(11):1777–1791
59. Matsuzaki K, Kato K, Yanagisawa K (2010) Abeta polymerization through interaction with membrane gangliosides. Biochim Biophys Acta 1801(8):868–877
60. Yanagisawa M, Yoshimura S, Yu RK (2011) Expression of GD2 and GD3 gangliosides in human embryonic neural stem cells. ASN Neuro 3(2):e00054. https://doi.org/10.1042/AN20110006
61. Yu RK, Suzuki Y, Yanagisawa M (2010) Membrane glycolipids in stem cells. FEBS Lett 584(9):1694–1699
62. Ogden AT, Waziri AE, Lochhead RA, Fusco D, Lopez K, Ellis JA, Kang J, Assanah M, McKhann GM, Sisti MB, McCormick PC, Canoll P, Bruce JN (2008) Identification of A2B5+CD133- tumor-initiating cells in adult human gliomas. Neurosurgery 62(2):505–514. discussion 514–505
63. Tchoghandjian A, Baeza N, Colin C, Cayre M, Metellus P, Beclin C, Ouafik L, Figarella-Branger D (2010) A2B5 cells from human glioblastoma have cancer stem cell properties. Brain Pathol 20(1):211–221
64. Sperling P, Franke S, Luthje S, Heinz E (2005) Are glucocerebrosides the predominant sphingolipids in plant plasma membranes? Plant Physiol Biochem 43 (12):1031–1038
65. Markham JE, Li J, Cahoon EB, Jaworski JG (2006) Separation and identification of major plant sphingolipid classes from leaves. J Biol Chem 281(32):22684–22694
66. Mamode Cassim A, Gouguet P, Gronnier J, Laurent N, Germain V, Grison M, Boutté Y, Gerbeau-Pissot P, Simon-Plas F, Mongrand S (2019) Plant lipids: key players of plasma membrane organization and function. Prog Lipid Res 73:1–27
67. Bayer EM, Mongrand S, Tilsner J (2014) Specialized membrane domains of plasmodesmata, plant intercellular nanopores. Front Plant Sci 5:507
68. Uemura M, Joseph RA, Steponkus PL (1995) Cold acclimation of *Arabidopsis thaliana* (Effect on plasma membrane lipid composition and freeze-induced lesions). Plant Physiol 109(1):15–30
69. Ischebeck T, Valledor L, Lyon D, Gingl S, Nagler M, Meijón M, Egelhofer V, Weckwerth W (2014) Comprehensive cell-specific protein analysis in early and late pollen development from diploid microsporocytes to pollen tube growth. Mol Cell Proteomics 13 (1):295–310
70. Mina JG, Okada Y, Wansadhipathi-Kannangara NK, Pratt S, Shams-Eldin H, Schwarz RT, Steel PG, Fawcett T, Denny PW (2010) Functional analyses of differentially expressed isoforms of the *Arabidopsis* inositol phosphorylceramide synthase. Plant Mol Biol 73(4–5):399–407
71. Rennie EA, Ebert B, Miles GP, Cahoon RE, Christiansen KM, Stonebloom S, Khatab H, Twell D, Petzold CJ, Adams PD, Dupree P, Heazlewood JL, Cahoon EB, Scheller HV (2014) Identification of a sphingolipid alpha-glucuronosyltransferase that is essential for pollen function in *Arabidopsis*. Plant Cell 26 (8):3314–3325
72. Msanne J, Chen M, Luttgeharm KD, Bradley AM, Mays ES, Paper JM, Boyle DL, Cahoon RE, Schrick K, Cahoon EB (2015) Glucosylceramides are critical for cell-type differentiation and organogenesis, but not for cell viability in *Arabidopsis*. Plant J 84 (1):188–201
73. Wang W, Yang X, Tangchaiburana S, Ndeh R, Markham JE, Tsegaye Y, Dunn TM, Wang GL, Bellizzi M, Parsons JF, Morrissey D, Bravo JE, Lynch DV, Xiao S (2008) An inositolphosphorylceramide synthase is involved in regulation of plant programmed cell death associated with defense in *Arabidopsis*. Plant Cell 20(11):3163–3179

74. Van Meer G, Lisman Q (2002) Sphingolipid transport: rafts and translocators. J Biol Chem 277(29):25855–25858
75. Peskan T, Westermann M, Oelmuller R (2000) Identification of low-density triton X-100-insoluble plasma membrane microdomains in higher plants. Eur J Biochem 267 (24):6989–6995
76. Kierszniowska S, Seiwert B, Schulze WX (2009) Definition of *Arabidopsis* sterol-rich membrane microdomains by differential treatment with methyl-beta-cyclodextrin and quantitative proteomics. Mol Cell Proteomics 8(4):612–623
77. Berkey R, Bendigeri D, Xiao S (2012) Sphingolipids and plant defense/disease: the "death" connection and beyond. Front Plant Sci 3:68
78. Begum MA, Shi XX, Tan Y, Zhou WW, Hannun Y, Obeid L, Mao C, Zhu ZR (2016) Molecular characterization of Rice OsLCB2a1 gene and functional analysis of its role in insect resistance. Front Plant Sci 7:1789
79. Ali U, Li H, Wang X, Guo L (2018) Emerging roles of sphingolipid signaling in plant response to biotic and abiotic stresses. Mol Plant 11(11):1328–1343
80. Nimrichter L, Rodrigues ML (2011) Fungal glucosylceramides: from structural components to biologically active targets of new antimicrobials. Front Microbiol 2:212
81. Rhome R, Del Poeta M (2010) Sphingolipid signaling in fungal pathogens. Adv Exp Med Biol 688:232–237
82. Insenser M, Nombela C, Molero G, Gil C (2006) Proteomic analysis of detergent-resistant membranes from *Candida albicans*. Proteomics 6(Suppl 1):S74–S81
83. Sonnino S, Mauri L, Chigorno V, Prinetti A (2007) Gangliosides as components of lipid membrane domains. Glycobiology 17 (1):1R–13R
84. Rodrigues ML, Nimrichter L, Oliveira DL, Frases S, Miranda K, Zaragoza O, Alvarez M, Nakouzi A, Feldmesser M, Casadevall A (2007) Vesicular polysaccharide export in *Cryptococcus neoformans* is a eukaryotic solution to the problem of fungal trans-cell wall transport. Eukaryot Cell 6(1):48–59
85. Rittershaus PC, Kechichian TB, Allegood JC, Merrill AH Jr, Hennig M, Luberto C, Del Poeta M (2006) Glucosylceramide synthase is an essential regulator of pathogenicity of *Cryptococcus neoformans*. J Clin Invest 116 (6):1651–1659
86. Ramamoorthy V, Cahoon EB, Thokala M, Kaur J, Li J, Shah DM (2009) Sphingolipid C-9 methyltransferases are important for growth and virulence but not for sensitivity to antifungal plant defensins in *Fusarium graminearum*. Eukaryot Cell 8(2):217–229
87. Zhu C, Wang M, Wang W, Ruan R, Ma H, Mao C, Li H (2014) Glucosylceramides are required for mycelial growth and full virulence in *Penicillium digitatum*. Biochem Biophys Res Commun 455(3–4):165–171
88. Oura T, Kajiwara S (2008) Disruption of the sphingolipid Delta8-desaturase gene causes a delay in morphological changes in *Candida albicans*. Microbiology 154 (Pt 12):3795–3803
89. Oura T, Kajiwara S (2010) *Candida albicans* sphingolipid C9-methyltransferase is involved in hyphal elongation. Microbiology 156 (Pt 4):1234–1243
90. Raj S, Nazemidashtarjandi S, Kim J, Joffe L, Zhang X, Singh A, Mor V, Desmarini D, Djordjevic J, Raleigh DP, Rodrigues ML, London E, Del Poeta M, Farnoud AM (2017) Changes in glucosylceramide structure affect virulence and membrane biophysical properties of *Cryptococcus neoformans*. Biochim Biophys Acta Biomembr 1859 (11):2224–2233
91. Lattif AA, Mukherjee PK, Chandra J, Roth MR, Welti R, Rouabhia M, Ghannoum MA (2011) Lipidomics of *Candida albicans* biofilms reveals phase-dependent production of phospholipid molecular classes and role for lipid rafts in biofilm formation. Microbiology 157(Pt 11):3232–3242
92. Perdoni F, Signorelli P, Cirasola D, Caretti A, Galimberti V, Biggiogera M, Gasco P, Musicanti C, Morace G, Borghi E (2015) Antifungal activity of Myriocin on clinically relevant *Aspergillus fumigatus* strains producing biofilm. BMC Microbiol 15:248
93. Cerantola V, Guillas I, Roubaty C, Vionnet C, Uldry D, Knudsen J, Conzelmann A (2009) Aureobasidin A arrests growth of yeast cells through both ceramide intoxication and deprivation of essential inositolphosphorylceramides. Mol Microbiol 71(6):1523–1537
94. Tan HW, Tay ST (2013) The inhibitory effects of aureobasidin A on *Candida* planktonic and biofilm cells. Mycoses 56 (2):150–156
95. Levery SB, Momany M, Lindsey R, Toledo MS, Shayman JA, Fuller M, Brooks K, Doong RL, Straus AH, Takahashi HK (2002) Disruption of the glucosylceramide biosynthetic pathway in *Aspergillus nidulans* and *Aspergillus fumigatus* by inhibitors of UDP-Glc:ceramide glucosyltransferase strongly affects spore germination, cell cycle,

and hyphal growth. FEBS Lett 525 (1–3):59–64

96. Thevissen K, Warnecke DC, François IE, Leipelt M, Heinz E, Ott C, Zähringer U, Thomma BP, Ferket KK, Cammue BP (2004) Defensins from insects and plants interact with fungal glucosylceramides. J Biol Chem 279(6):3900–3905

97. Thevissen K, de Mello TP, Xu D, Blankenship J, Vandenbosch D, Idkowiak-Baldys J, Govaert G, Bink A, Rozental S, de Groot PW, Davis TR, Kumamoto CA, Vargas G, Nimrichter L, Coenye T, Mitchell A, Roemer T, Hannun YA, Cammue BP (2012) The plant defensin RsAFP2 induces cell wall stress, septin mislocalization and accumulation of ceramides in *Candida albicans*. Mol Microbiol 84(1):166–180

98. Tavares PM, Thevissen K, Cammue BP, François IE, Barreto-Bergter E, Taborda CP, Marques AF, Rodrigues ML, Nimrichter L (2008) *In vitro* activity of the antifungal plant defensin RsAFP2 against *Candida* isolates and its *in vivo* efficacy in prophylactic murine models of candidiasis. Antimicrob Agents Chemother 52(12):4522–4525

99. Rodrigues ML, Travassos LR, Miranda KR, Franzen AJ, Rozental S, de Souza W, Alviano CS, Barreto-Bergter E (2000) Human antibodies against a purified glucosylceramide from *Cryptococcus neoformans* inhibit cell budding and fungal growth. Infect Immun 68(12):7049–7060

100. Rodrigues ML, Shi L, Barreto-Bergter E, Nimrichter L, Farias SE, Rodrigues EG, Travassos LR, Nosanchuk JD (2007) Monoclonal antibody to fungal glucosylceramide protects mice against lethal *Cryptococcus neoformans* infection. Clin Vaccine Immunol 14 (10):1372–1376

101. Xisto MIDDS, Henao JEM, Dias LDS, Santos GMP, Calixto ROR, Bernardino MC, Taborda CP, Barreto-Bergter E (2019) Glucosylceramides from *Lomentospora prolificans* induce a differential production of cytokines and increases the microbicidal activity of macrophages. Front Microbiol 10:554

102. Da Silva AFC, Rodrigues ML, Farias SE, Almeida IC, Pinto MR, Barreto-Bergter E (2004) Glucosylceramides in *Colletotrichum gloeosporioides* are involved in the differentiation of conidia into mycelial cells. FEBS Lett 561(1–3):137–143

103. Nimrichter L, Barreto-Bergter E, Mendonça-Filho RR, Kneipp LF, Mazzi MT, Salve P, Farias SE, Wait R, Alviano CS, Rodrigues ML (2004) A monoclonal antibody to glucosylceramide inhibits the growth of *Fonsecaea pedrosoi* and enhances the antifungal action of mouse macrophages. Microb Infect 6(7):657–665

104. Pinto MR, Rodrigues ML, Travassos LR, Haido RM, Wait R, Barreto-Bergter E (2002) Characterization of glucosylceramides in *Pseudallescheria boydii* and their involvement in fungal differentiation. Glycobiology 12(4):251–260

105. Rollin-Pinheiro R, Liporagi-Lopes LC, De Meirelles JV, Souza LM, Barreto-Bergter E (2014) Characterization of *Scedosporium apiospermum* glucosylceramides and their involvement in fungal development and macrophage functions. PLoS One 9(5):e98149

106. Mor V, Rella A, Farnoud AM, Singh A, Munshi M, Bryan A, Naseem S, Konopka JB, Ojima I, Bullesbach E, Ashbaugh A, Linke MJ, Cushion M, Collins M, Ananthula HK, Sallans L, Desai PB, Wiederhold NP, Fothergill AW, Kirkpatrick WR, Patterson T, Wong LH, Sinha S, Giaever G, Nislow C, Flaherty P, Pan X, Cesar GV, de Melo TP, Frases S, Miranda K, Rodrigues ML, Luberto C, Nimrichter L, Del Poeta M (2015) Identification of a new class of antifungals targeting the synthesis of fungal sphingolipids. MBio 6(3):e00647

107. Lazzarini C, Haranahalli K, Rieger R, Ananthula HK, Desai PB, Ashbaugh A, Linke MJ, Cushion MT, Ruzsicska B, Haley J, Ojima I, Del Poeta M (2018) Acylhydrazones as antifungal agents targeting the synthesis of fungal sphingolipids. Antimicrob Agents Chemother 62(5):e00156-18. https://doi.org/10.1128/AAC.00156-18

108. Siebers M, Brands M, Wewer V, Duan Y, Holzl G, Dormann P (2016) Lipids in plant-microbe interactions. Biochim Biophys Acta 1861(9 Pt B):1379–1395

109. Koga J, Yamauchi T, Shimura M, Ogawa N, Oshima K, Umemura K, Kikuchi M, Ogasawara N (1998) Cerebrosides A and C, sphingolipid elicitors of hypersensitive cell death and phytoalexin accumulation in rice plants. J Biol Chem 273(48):31985–31991

110. Umemura K, Ogawa N, Yamauchi T, Iwata M, Shimura M, Koga J (2000) Cerebroside elicitors found in diverse phytopathogens activate defense responses in rice plants. Plant Cell Physiol 41(6):676–683

111. Umemura K, Tanino S, Nagatsuka T, Koga J, Iwata M, Nagashima K, Amemiya Y (2004) Cerebroside elicitor confers resistance to fusarium disease in various plant species. Phytopathology 94(8):813–818

Chapter 4

Insights into Yeast Phospholipid Tra(ffi)cking

Malathi Srinivasan and Ram Rajasekharan

Abstract

Lipids are the best known form of energy storage. Apart from being energy reserves and signaling molecules, they also form an integral part of the cell membranes and provide structure and fluidity to the membranes. A group of polar lipids, namely, the phospholipids (PLs), are membrane constituents, defining the structure, shape, and function of the cells. They decide the cell permeability due to their hydrophobicity. PLs also play an under-rated role in many human diseases, due to which more importance needs to be given toward studies focusing on their synthesis, function, and metabolism. The budding yeast, *Saccharomyces cerevisiae*, is a model organism for the study of PLs as it is a simple system to characterize lipid metabolic changes under various physiological conditions. Yeasts have been used to study the mechanisms related to lipid metabolism, lipid trafficking, and localization in different subcellular organelles. It also shows the presence of various PLs, which makes it a versatile tool for research. With the recent developments in detection and quantification of PLs, and techniques using novel reagents, tags, and specific stains to localize a particular lipid in different subcellular compartments, yeast makes itself a remarkable model for lipid research. In this chapter, we discuss the various PLs present in the budding yeast, their traffic within the yeast cell, and methods of tracking them.

Keywords Phospholipids, Yeast, Biosynthesis, Tracking, Mass spectrometry, Enzyme assays

1 Introduction

Lipids are broadly defined as hydrophobic or amphiphilic small molecules and include fatty acids, phospholipids, sterols, monoacylglycerol, diacylglycerol, triacylglycerol, and sphingolipids [1]. While the triglycerides and steryl esters act as a storage form of energy and get stored as lipid droplets in yeast, the phospholipids are basic structural components of the membranes of the cell and its organelles. A neutral glycerolipid molecule consists of a glycerol backbone and fatty acid molecule(s). Examples are the mono-, di-, and triacylglycerols (MG, DG, and TGs). These are also called as the nonpolar or neutral lipids as they are strongly hydrophobic and carry no charge. Polar glycerolipids, on the other hand, are amphipathic and charged; they include the phospholipids and betaine glycerolipids [2]. Phospholipids are composed of three sections: a

Rajendra Prasad and Ashutosh Singh (eds.), *Analysis of Membrane Lipids*, Springer Protocols Handbooks,
https://doi.org/10.1007/978-1-0716-0631-5_4,

diacylglycerol, a hydrophilic "head" consisting of the charged phosphate moiety, and a small organic molecule that is covalently bound to the phosphate. Besides their role in membrane structure, lipids act as the key second messengers in intracellular signaling [3], the most important of them being the diacylglycerol (DG). It is to be noted that both DG and TG are formed from the precursor molecule, phosphatidic acid (PA), which is a phospholipid. PA is in fact an important cell lipid since it is the biosynthetic precursor for the formation, either directly or indirectly, of most acyl-glycerol lipids in the cell. PA also structurally impacts the membrane curvature and helps in recruiting cytosolic proteins to suitable membranes. The conversion of PA into DG by lipid phosphate phosphohydrolases (LPPs) is the commitment step for the production of other major phospholipids, namely, phosphatidylcholine (PC), phosphatidylethanolamine (PE), and phosphatidylserine (PS). Furthermore, the DG thus formed is also converted into cytidine diphosphate diacylglycerol (CDP-DAG), which is a central liponucleotide intermediate and the precursor for phosphatidylglycerol (PG), phosphatidylinositol (PI), and phosphoinositide (PIP, PIP_2, PIP_3). While PE is found in all living cells, composing nearly 25% of all phospholipids, they make up 45% of all phospholipids of the human brain, where they are found particularly in the white matter of brain, nerves, neural tissue, and in the spinal cord. PS is crucial in blood coagulation and also in cell signaling related to apoptosis. In 2003, the FDA has given a qualified health claim to PS that "consumption of phosphatidylserine may reduce the risk of dementia and cognitive dysfunction in the elderly" along with certain disclaimers on having little evidence [4]. Similarly, PG, which is a pulmonary surfactant constituent, plays a key role in the mitochondrial inner membrane as cardiolipin (CL). Two molecules of PG condense to form CL which is essential in the normal functioning of the enzymes of the mitochondrial membrane and energy homeostasis.

The homeostasis of glycerophospholipids forms an indispensable factor for build-up, maintenance, and the dynamics of cell membranes and hence their curvature in all organisms. The various phospholipid molecular species found in yeast basically vary in the composition of the fatty acids which are esterified to the sn-1 and sn-2 position of a phosphorylated glycerol, the phosphorylating head group being a choline, ethanolamine, L-serine, *myo*-inositol, or sometimes the phospholipid without a head group, i.e., phosphatidate [5].

Since they form the integral part of the membrane bilayer, their physical and chemical properties are critical to the formation and stability of the membranes which serve as permeability checkpoints to the hydrophilic substances like ions and metabolites. Apart from this, the phospholipid bilayers also provide suitable conditions for the biogenesis and metabolism of amphipathic biological molecules

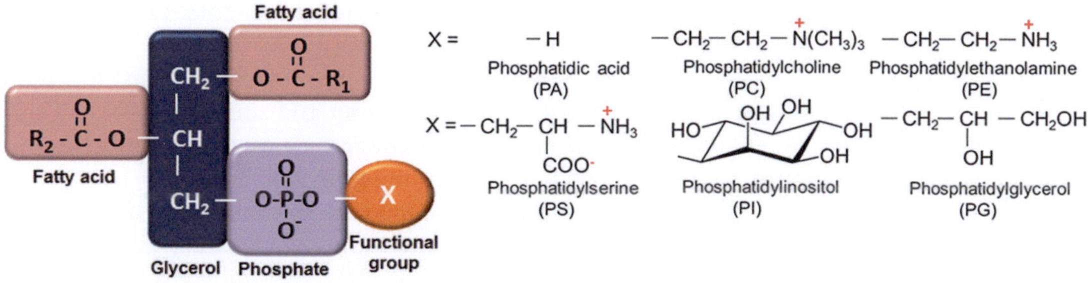

Fig. 1 A schematic representation of a phospholipid molecule. The functional group (X) defines the nature of phospholipids [7]

including lipids and membrane proteins. Thus, phospholipids play a very significant role in the structure and function of the cells, besides playing an under-rated role in many human diseases, due to which more importance needs to be given toward studies on their synthesis, function, and metabolism. Phospholipids could also serve as potential biomarkers for some pathological conditions. For example, there is growing evidence that they may be potentially involved in non-alcoholic fatty liver disease, atherosclerosis, viral infections, and cancer, via the activity of lyso-phosphatidylcholine acyl transferase (LPCAT) that catalyzes the recruitment of fatty acids into the *sn-2* position of PC [6] and recent research focuses on the pharmacological manipulation of LPCAT activity and membrane phospholipid composition as new therapeutic options for these conditions. This emphasizes the necessity to know more about phospholipids, both qualitatively and quantitatively (Fig. 1).

1.1 Biosynthesis and Hydrolysis of Phospholipids in Yeast

The budding yeast, *Saccharomyces cerevisiae,* serves as an excellent model organism for studying the physiological functions of multicellular eukaryotes like humans, because of their structural similarity at the subcellular level. It has long been known as a versatile tool for biotechnology and studies on lipid homeostasis. The unicellular nature of yeast makes it especially susceptible to environmental stress, and with evolution, yeast has developed elaborate signalling pathways to maintain lipid homeostasis [8]. The availability of the whole yeast genome sequence helps researchers to understand the homologies between yeast and higher eukaryotes. The simplicity of generation of deletion mutants in *S. cerevisiae* also helps researchers to understand different metabolic pathways such as lipid metabolism. Yeast cell metabolism is holistic in fulfilling its cellular lipid requirements by different pathways, including de novo synthesis, uptake of external lipids, and turnover of lipids [9]. Studies on yeast have helped in delimiting mechanisms related to lipid metabolism, lipid trafficking, and localization in different subcellular organelles at various growth stages. Therefore, to study the phospholipid metabolism, like for all other metabolic pathways, *S. cerevisiae* is a suitable experimental organism. Furthermore, the availability of modern tools and techniques using novel reagents, tags, and specific stains to localize a particular lipid in different subcellular compartments in yeast makes it a remarkable means for lipid research. Yeast provides an outstanding and simple system to characterize lipid metabolic changes under various physiological conditions [10] that may resemble human physiology.

PL synthesis in yeast is defined by two alternative pathways, namely, the CDP-DAG (cytidine diphosphate-diacylglycerol) and Kennedy pathways. It mainly occurs in ER while the major non-bilayer-forming phospholipids like PE and cardiolipin (CL) are synthesized in the mitochondria [11, 12]. Post de novo synthesis of phospholipids in the Kennedy pathway, they can be degraded and remodeled. Remodeling of PLs was proposed in the famous "Lands cycle" with a de-acylation-re-acylation process [13], typically characterized by saturated acyl chains at the *sn*-1 and unsaturated acyl chains at the *sn*-2 position [14]. PL hydrolysis occurs by both phospholipases and lipid phosphatases. There are four major types of phospholipases, A, B, C, and D; C and D types are classified as phosphodiesterases. Their roles include digestion, cell signaling, and disruption of organelle cell membranes [15].

The phospholipid composition, primarily the fatty acid that it constitutes of, decides the nature of the cell membranes; but all organelles of a cell cannot synthesize their membrane lipids, thereby necessitating a transfer from the site of synthesis to the cellular membrane(s) of destination. This further necessitates modern tools to track the phospholipids. The distribution of phospholipids among cellular membranes occurs in an organized fashion. Phospholipid remodeling by phospholipases is essential for the maintenance of the constant lipid composition of membranes. Unlike proteins, lipids do not possess signals for their distribution to the desired destination. Therefore, other not-well-known mechanisms may govern the distribution and intracellular transport of lipids from one hydrophobic compartment to another. A distinct pattern of proteins and lipids is found in each cellular membrane of eukaryotes [16]. Like marker proteins and enzymes, individual lipids are also characteristic for certain membranes. The lipid composition of different membranes varies throughout the cell. Table 1 shows the composition of different phospholipids in the organelles of yeast [9]. Given such qualitative and quantitative specificities, it becomes essential that there are methods to track and quantitate these phospholipids in various cell organelles.

1.2 Role of Phospholipids

Besides acting as a barrier, phospholipids also play an important role in budding, intracellular membrane trafficking, fission, and fusion in yeast. Some of the major yeast phospholipids are explained in this chapter, along with methods available for tracking them inside the yeast cell.

Table 1
Phospholipid composition of organelles from *S. cerevisiae*

	Mol% of total phospholipids						
Cell fraction	**PC**	**PE**	**PI**	**PS**	**CL**	**PA**	**Others**
Homogenate	51.0	25.0	11.4	5.1	3.7	1.1	2.7
Plasma membrane	11.3	24.6	27.2	32.2	nd	3.3	1.4
Endoplasmic reticulum	38.9	18.6	22.4	6.4	0.3	3.4	10.0
Mitochondria	33.4	22.7	20.6	3.3	7.2	1.7	10.1
Peroxisomes	39.8	17.4	22.0	2.5	2.7	6.1	10.5

1.2.1 Phosphatidyl Glycerol

As mentioned, PG is not a major membrane PL. Being negatively charged, phosphatidylglycerol introduces the negative charge onto the membrane surface and to the lipid–protein interface. The phosphatidylglycerol acyl chains assume particular conformations on the protein hydrophobic transmembrane surface [17]. PG is synthesized from phosphatidylglycerol phosphate (PGP), which in turn is synthesized by PGP synthase (Pgs1p) which uses CDP-DG and G3P [18]. This is followed by a phosphatase reaction where the phosphate group is removed by Gep4p, resulting in PG [19]. This can be further converted to Cardiolipin (CL) by the cardiolipin synthase Crd1p [20, 21].

The major role of PG is to act as a precursor for mitochondrial cardiolipin. Unlike other phospholipids, cardiolipin is unique in structure as well as localization. CL was first isolated from the beef heart [22]. Although CL became the first-characterized phospholipid, it was the last of the major phospholipids to be stereochemically defined [23]. CL (1,3 diphosphatidyl-*sn*-glycerol) is a unique anionic phospholipid with a dimeric structure having four fatty acyl chains (Fig. 2). CL is almost exclusively localized to the mitochondrial inner membrane in eukaryotes [22, 23].

1.2.2 Role of Cardiolipin

CL is a signature phospholipid species of the mitochondria. The mitochondria are crucial organelles that control the life and death of the cell. Important metabolic reactions, the synthesis of most of the cellular ATP, and the regulation of a number of signaling cascades, such as apoptosis, occur in the mitochondria [24]. The major non-bilayer-forming phospholipids PE and CL are synthesized in the mitochondria [11, 12]. The presence of CL and PE is important for mitochondrial fusion [25]. CL is needed for the function and stability of several mitochondrial protein complexes, such as the ADP/ATP carrier and respiratory chain complex III and IV [26, 27]. In the absence of CL, a reduction in the inner membrane potential $\Delta\psi$ and a defect in protein import into the

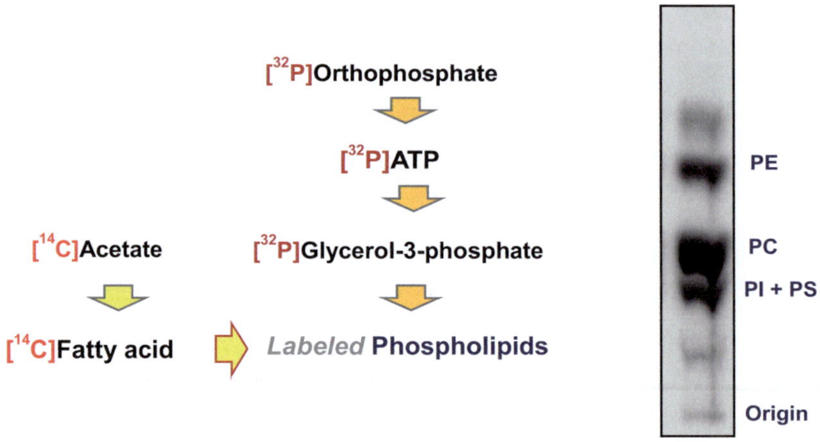

Fig. 2 Radiolabeling of yeast phospholipids using [^{32}P]orthophosphate or [^{14}C]acetate

mitochondria have been reported [28]. It has been suggested that CL and mitochondrial PE have overlapping functions and that PE can compensate for the loss of CL and vice versa [29]. It is well established that CL remodeling can be achieved via two pathways, similar to the remodeling of other phospholipids. The first occurs via the deacylation–reacylation cycle [coenzyme A (CoA)-dependent] reported by Schlame and Rustow [30], and the second occurs via a deacylation–transacylation cycle [CoA-independent] between acceptor and donor phospholipids, as proposed by Xu et al. [31]. Barth syndrome patients have deficiencies in molecular species of CL, PG, PC, and PE [32, 33]. Several models based on the deacylation-transacylation cycle have been proposed for CL remodeling in yeast [34–36]. PG is a precursor of CL biosynthesis [28]. Recently, Pokorna et al. [37] have reported the presence of two diverse pools of PG with different fatty acid compositions in the yeast mitochondria. Like CL remodeling, PG remodeling has been reported in human mitochondria [38], but there is no report in yeast. Several studies have shown that CL is very important for mitochondrial function and growth on different carbon sources [19, 34, 36]. The growth defect is more pronounced at higher temperatures because CL is required to maintain the mitochondrial DNA and other important cellular functions [28]. Defects in PE and CL metabolism cause mitochondrial dysfunction.

Lee et al. [39] reported that mitochondrial dysfunction causes the formation of lipid droplets (LDs) in the cells as a general response to stress. Eukaryotic cells store lipids mostly as TG and steryl esters in a specialized organelle of the cell called a lipid droplet, lipid particle, lipid body, or oil body [40]. LDs consist of a core of nonpolar lipids covered with a monolayer of phospholipids that have peripheral and embedded proteins [41–43]. LDs are dynamic organelles and play a key role in the storage and mobilization of nonpolar lipids. Despite their crucial role in lipid metabolism and energy homeostasis, the cellular biology of LDs and its metabolic link with other organelles are poorly understood. Horvath et al. [44] reported a physiological link between PE and TG metabolism in *S. cerevisiae*, involving an acyl-CoA-independent pathway through the phospholipid:diacylglycerol acyltransferase (PDAT) activity of the Lro1 protein. However, the role of CL in TG metabolism or the formation of LDs remains unknown. Several studies have shown that LDs are physically connected to the mitochondria [45–47], suggesting that CL might affect TG metabolism. In yeast, lipid biosynthesis is a growth phase-dependent process [44]. During the exponential growth phase, phospholipid synthesis mainly occurs for cell growth, membrane, and organelle formation. Nonpolar lipid (TG and SE) synthesis and the accumulation in cytosolic LDs are characteristics of the stationary growth phase. The CL biosynthetic genes are up-regulated in the stationary-phase cells [36], suggesting that CL is the only phospholipid whose synthesis increases in the stationary phase. These findings suggest a physiological link between CL and TG metabolism.

Tracking of Cardiolipin: CL was traditionally quantified using a dye, 10-*N*-nonyl acridine orange (NAO), which was supposed to specifically bind to cardiolipin [48]. However, it was observed that the dye also binds to certain nonspecific mono acidic phospholipids, which necessitated the use of a fluorimetric method, based on the red fluorescence of the dye dimers formed at the diacidic phospholipid contact [49]. Although early data suggested that the mitochondrial uptake of NAO was independent of membrane potential, later studies by Jacobson et al. [50] reported otherwise. Likewise, the high affinity of NAO for CL which was thought to result from an electrostatic and a hydrophobic interaction [48] was later found to be due to the insertion of the nonyl chain into the bilayer at the hydrophobic surface of the membrane [51].

In a study reported in 2015 [52], Morita and Terada have developed a novel fluorimetric method to measure both PG and CL using combinations of specific enzymes and Amplex Red. This assay estimated the combined amount of PG and CL (PG + CL) irrespective of the species of the fatty acid side chains that were attached. There are also fluorimetric kits that are available to quantify the amount of cardiolipin in cell lysates or isolated mitochondria. These kit manufacturers promote easy detection using these probes as compared to mass spectrometry which has been employed in the last couple of decades for the detection of lipids, as it requires the use of costly, sophisticated instrumentation, and expertise.

1.2.3 Phosphatidylcholine

The main route of PC synthesis in yeast, which does not operate in higher plants, is the one that happens in the liver of higher animals. It is the SAM (*S*-adenosylmethionine)-mediated pathway, the lesser known pathway being that from phosphocholine. SAM is the source of methyl groups for the sequential methylation of PE to the intermediary mono and dimethyl PE. This reaction is catalyzed by *N*-methyltransferases. At least two isoforms of this enzyme are reported in yeast (Cho2/Pem1 and Opi3/Pem2). While PC is a major lipid in yeasts, recent work suggests that it is not as crucial as it is in higher organisms, especially if a suitable alternative growth substrate is available, although reports say that a depletion on PC in yeast results in shortening and increased saturation of the fatty acyl chains, thereby regulating the intrinsic membrane curvature [53]. Similar effects on peroxisome membranes were also reported when PC synthesis was disrupted in certain mutants [54]. Yeast uses the enzyme glycerophosphocholine acyltransferase (Gpc1) to convert glycerophosphocholine (formed from PC) to lysophosphatidylcholine (LPC), which can be recycled to PC by the lysophospholipid acyltransferase (Ale1), while significantly changing the molecular species as compared to the parent PC. This process is precisely coordinated, and it plays a significant role in the growth of the organism. In the PC remodeling pathway, the de-acylation of 1,2-dipalmitoyl-*sn*-glycero-3-phosphocholine is catalyzed by the phospholipase A2 activity of the triacylglycerol lipase TGL4p [55]. As for its location, PC is present in a non-covalently bound form in the membrane proteins such as the yeast cytochrome bc1. Since PC is a very dominant phospholipid species in the membranes, its homeostasis must be under a very rigid control in order to ensure the proper functioning of cell membranes [53, 56].

Tracking of PC: PC can be readily isolated and analyzed by TLC and HPLC, without much problem. Determination of the dipalmitoyl species is more strenuous, but specific methods using modern mass spectrometry have addressed this. In studies which involve the position of the fatty acids on the glycerol moiety, phospholipase A_2 from snake venom is used. The hydrolyzed product lysophosphatidylcholine is analyzed in these studies. However, the inadvertent formation of LPC can impede such quantifications.

1.2.4 Phosphatidylethanolamine

PE represents the backbone of most biological membranes [57] and plays a role in membrane fusion, membrane curvature, and cell division. It forms an important substrate in several important biopathways. Compared to PC, the polar head group of PE creates a more viscous lipid membrane. It has a lower melting point than PC, resulting in more fluid membranes. Sequential methylation of PE results in mono and dimethyl-PE, which also serves as a route for PC biosynthesis, especially in yeast and bacteria. PE is essential for the growth of yeast cells and is synthesized through any of the four different enzymatic pathways which include the decarboxylation of

PS, at the mitochondrial membrane via phosphatidylserine decarboxylase 1 (Psd1), or through phosphatidylserine decarboxylase 2 (Psd2) at the Golgi and vacuolar membranes. It can also be possibly produced from actively retrieved extracellular ethanolamine through the cytidyldiphosphate ethanolamine (CDP-Etn) branch of the Kennedy pathway, or by a seldom followed pathway which is based on the lysophospholipid acylation of lyso-PE. Unlike other PLs, PE is not recruited spontaneously into the bilayers but instead incorporates into curved structures [57]. Additionally, in yeast, PE also functions as an anchor to autophagosomal membranes for the autophagy-related protein Atg8. Similar to CL, PE is also linked to TG metabolism in yeast. According to Horvath et al. [44], TG and PE syntheses in yeast are tightly linked, since the TG formation by the acyltransferase Lro1p is highly dependent on PE synthesis. Lro1p preferentially uses PE as co-substrate to transfer its *sn-2* acyl group to DG, resulting in the formation of TG and lyso-PE. Thus, the PE amounts in the cell influence the TG synthesis. Several studies from the laboratory of Gunther Daum using appropriate mutants have looked into the PE contribution to TG synthesis through various PE biosynthetic pathways until it was made evident that there exists a metabolic link between TG formation and PE synthesis via the CDP-Etn pathway but not by the other three PE biosynthetic routes [44].

Tracking of PE: PE is one of the glycerophospholipids that can easily be analyzed using thin-layer chromatographic methods. Two-dimensional thin-layer chromatography on silica gel plates using chloroform/methanol/25% NH_3 (68:35:5; per vol) as the first and chloroform/acetone/methanol/acetic acid/water (53:20:10:10:5; per vol) as the second developing solvent system is usually used following which the PE can be visualized by staining with iodine vapor, scraped off the plate, and quantified using the method of Broekhuyse [58]. Due to its abundance, PE can be visually observed under iodine vapor. Accurate quantification can be done using mass spectrometry.

1.2.5 Phosphatidylserine

Phosphatidylserine (PS) is composed of PA, with the negatively charged phosphate group attached to the amino acid serine at the hydroxyl end. The presence of the amino acid serine in addition to the negative phosphate group confers a net negative charge on PS. PS is produced by the addition of serine to cytidine diphosphate diacylglycerol (CDP-DAG) by PS synthase in bacteria and yeast. In addition to its function as a membrane component and as a precursor for other major phospholipids, PS serves as an essential cofactor in binding and activating a large number of proteins, significantly of those involved in cell signaling. The negative charge on the PS further helps in the binding through electrostatic interactions or Ca^{2+} bridges. It is also often referred to as a cellular marker for apoptosis, or programmed cell death [59]. Though a membrane PL, PS normally occupies the inner leaflet of the membrane and any deviation from this location, making it appear on the cell surface is a marker for apoptosis and blood clotting. Another lesser known function of PS is in pathogenesis and virulence. PS and PE have been less studied compared to some other phospholipid classes with regard to their roles in virulence. However, recent studies reveal interesting roles for PS in the virulence of *Candida albicans* as well as a variety of protozoan and prokaryotic pathogens [60]. In spite of these important functions, PS is often underrated and not much is reported about its subcellular localization and intracellular dynamics.

Tracking of PS: Since PS is an acidic lipid, the metal ions associated with it interrupt analyses. Two-dimensional TLC can be used for separation just like with other PLs, although HPLC has not been a very effective separation method. One of the early methods to detect PS involved a covalent binding of 2,4,6-trinitrobenzenesulfonate (TNBS) with the amine.

Recent studies, however, employ mass spectrometry for molecular species analysis and quantification. There are also genetically encoded probes for detection of PS within live cells. Use of antibodies and a probe like Annexin have also been reported. The risk here, however, is the nonspecific binding with other phospholipids. Methods of using 7-nitro-2-1-3-benzoxadiazol-4-yl (NBD) that helps in getting fluorescently labeled PS analogs have not been very successful as the NBD-PS results were inconclusive due to the rapid bleaching and the rapid removal of the analog from the membranes. Another fluorophore that has been used is BODIPY. More recent studies have shown the use of a fluorescent probe construct that links the lactadherin C2 domain with the green fluorescent protein (GFP-LactC2) that can be easily transfected into yeast cells, which provides very good results in PS trafficking [61].

1.2.6 Phosphatidylinositol

A *myo*-inositol-containing phospholipid, PI, is the third most abundant PL in the membrane of *S. cerevisiae*. PI is synthesized from CDP-DAG and *myo*-inositol by PI synthase (PIS). PIS, unlike the other yeast phospholipid-synthesizing enzymes, is a constitutive enzyme, whose expression is unaffected by the addition of *myo*-inositol and choline to culture medium or the transition of growth phase. *PIS* locus deletion in the genome is lethal, which shows that PI is essential for the survival and growth of yeast cells [62]. PI and its phosphorylated products (phosphatidylinositol phosphates—PIPs) are involved in various functions like signal transduction, mRNA export, vesicle trafficking, and glycosylphosphatidylinositol synthesis. Vesicular transport is enabled via the recruitment of certain proteins, by the phosphorylation and dephosphorylation of the inositol headgroups of the PIPs at specific loci on the membrane [63]. Protein recruitment is based on the presence of a PIP-specific binding domain. One of the well-studied domains is the pleckstrin homology (PH) domain. Numerous conserved kinases and phosphatases catalyze the synthesis and interconversion of the PIPs; these kinases are highly specific and will phosphorylate only one substrate. The yeast genome encodes three kinase (PI4K) isoforms, viz., Pik1, Stt4, and Lsb6; while Pik1 is cytosolic, the latter two are membrane bound. Studies have shown that although both Pik1 and Stt4 catalyze the same biochemical reaction, they play nonoverlapping roles [64], such that the overexpression of Stt4 cannot rescue the phenotype of Pik1 deletion mutant; deletion of Stt4 in yeast is lethal. The role of phosphoinositides in signaling has been a topic of research interest for the last several years. As much as the synthesis of the PIPs is important, their degradation has also received research attention, While Inp51 is the major phosphatase, Sac1, located on the ER and Golgi is also an important phosphatase; Sac1 can counteract Pik1 activity [65].

Tracking of PI: Phosphoinositides play key roles in cell function, and hence, detection and estimation and localization of these metabolites are indispensable. Hence, several modern methods are developed for tracking them. Early biochemical detection methods included radiolabeling, as PIs are minor PLs and high sensitive methods were needed. However, the traditional TLC and HPLC coupled with mass spectrometry which not only analyzes the head group but also detects the fatty acyl chains are getting more popular, due to its sensitivity and accuracy. But biochemical methods have the limitation of poor temporal resolution, where the PI composition is detected but not the changing levels of the lipid. This is where imaging mass spectrometry comes to play. Other microscopic methods such as use of fluorescent probes, use of protein modules that employ PI binding module-based biosensors, total internal reflection fluorescence (TIRF) microscopy, and super-resolution video microscopy are being reported in the recent years [66].

1.2.7 Methods of Tracking Phospholipids

1. *Thin-layer chromatography*: Early methods of PL tracking depended on chromatographic separation using thin-layer chromatography. Lipid extraction is done following the method of [67]. Yeast cells grown to logarithmic phase ($A_{600} = 20$) are harvested, and lipid is extracted using chloroform:methanol:acidified water (1:1:1 v/v/v). The extracted lipids are then concentrated and resolved using a silica-TLC plate. Since PLs are polar lipids, a water-based solvent system like chloroform:acetone:methanol:acetic acid: water (50:20:10:15:5 v/v is used).

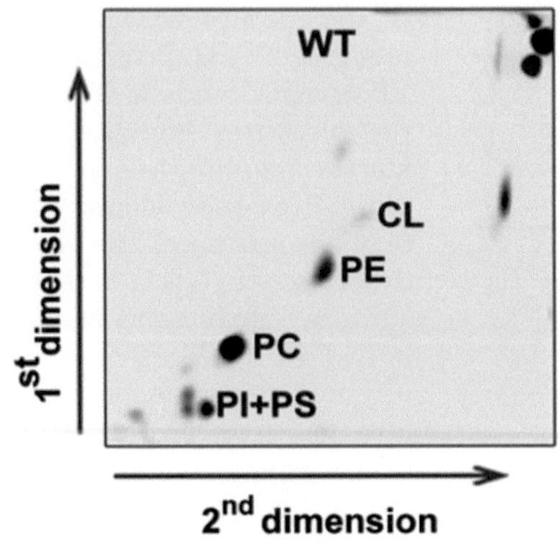

Fig. 3 [^{14}C]Acetate labeled yeast phospholipids, resolved using two-dimensional chromatography. *CL* cardiolipin, *PC* phosphatidylcholine, *PE* phosphatidylethanolamine, *PI* phosphatidylinositol, *PS* phosphatidylserine, *WT* wild type

To enable more accurate tracking and quantification of the PL species, radiolabeling methods have been used. The yeast cells are grown in a medium containing either 0.2 μCi/ml of [^{14}C] acetate or 2 μCi/ml [^{32}P] phosphate radioisotope. Post the lipid extraction and resolution on TLC as described above, the plates are viewed and imaged under a phosphor imager. Distinct bands pertaining to various PL species can be obtained, and the same can be quantified either using an ImageJ software or can be more accurately calculated using a liquid scintillation counter by scraping the silica pertaining to the various bands and dissolving it in a scintillation fluid. Caution must be exercised while handling radioisotopes.

For more accurate and finer resolution of the individual PLs, two-dimensional thin-layer chromatography is done. The extracted lipids are separated in the first dimension using a primary solvent system such as chloroform:methanol:acetic acid:water, 85:15:10:3.5, v/v. This helps in the separation of PA, PC, PE, PI, PS, CL, and PG. To further separate the CL and PG from the other phospholipids, the TLC plate is placed in the second solvent system, consisting of chloroform:methanol:glacial acetic acid, 65:25:8, v/v.

Alternately, a different set of solvent system can also be used to run the two-dimensional chromatography. This includes a basic solvent system followed by an acidic system. The first dimension uses chloroform:methanol:25% ammonia (65:35:4, v/v), and the second dimension uses chloroform: methanol:acetic acid:water (85:25:5:4, v/v) (Fig. 3).

Cardiolipin hydrolyzes to form Monolysocardiolipin (MLCL). This can also be detected using TLC by the following method. The extracted lipids are resolved using chloroform: methanol:acetic acid:water, 85:25:5:4, v/v, as the solvent system, and the plate is subsequently charred with 10% cupric sulfate in 8% aqueous phosphoric acid solution for 10 min at 180 °C. The intensity of the charred bands can be quantified using an ImageJ software or densitometry.

2. *Enzymatic assays*: Since most PLs form specific substrates for various acyltransferases and lipases in the synthesis and hydrolysis of the PLs, respectively, enzymatic assays can also be used as a tool for tracking the PLs within a cell organelle. Microsomal preparations and isolated mitochondria can serve as the source of enzymes. Commercially available PLs—either radiolabeled or fluorescent tagged—can be used as substrates.

   (i) Example of lysophospholipid acyltransferase assays:

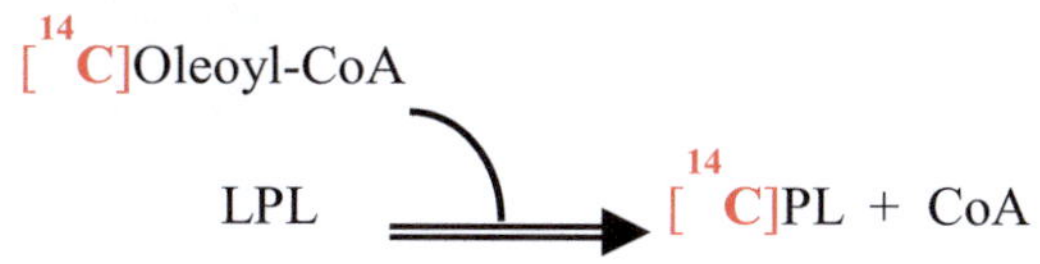

LPL - Lysophospholipid; PL - Phospholipid

   (ii) Example of cardiolipin hydrolysis:

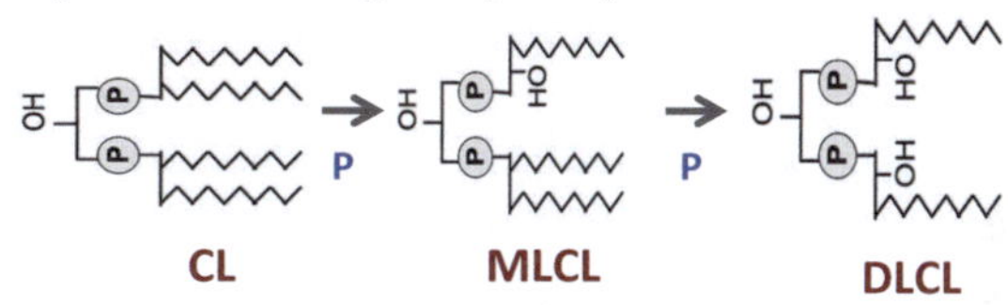

Some of the commonly used fluorescent substrates are TopFluor-CL, 1,1′,2,2′-tetraoleoyl cardiolipin[4-(dipyrromethene boron difluoride)butanoyl], NBD-PA, 1-oleoyl-2-{12-[(7-nitro-2–1,3-benzoxadiazol-4 yl)amino]dodecanoyl}-*sn*-glycero-3-phosphate, NBD-PC, 1-oleoyl-2-[12-[(7-nitro-2–1,3-benzoxadiazol-4-yl)amino]dodecanoyl]-sn-glycero-3-phosphocholine, NBD-PG, 1-oleoyl-2-{12-[(7-nitro-2–1,3-benzoxadiazol-4-yl)amino]dodecanoyl}-*sn*-glycero-3-[phosphorac-(1-glycerol)], NBD-PS, 1-oleoyl-2-{12-[(7-nitro-2–1,3-benzoxadiazol-4-yl)amino]dodecanoyl}-*sn*-glycero-3-phosphoserine)], and NBD-PE (*N*-(NBD-aminohexanoyl)-1,2-dioleoyl-*sn*-glycero-3-phosphoethanolamine. Hydrolysis of these substrates by the lipases results in the formation of fluorescently tagged lysolipid, which can be further quantified. In short, the assay is conducted as follows: The assay mixture consists of 50 mM Tris, pH 8.0, 3 mM $MgCl_2$, 1 mM EDTA, 5 mM $CaCl_2$, 250 μM CHAPS, 40% glycerol, 5 μM fluorescently labeled substrate, and the enzyme protein in a total volume of

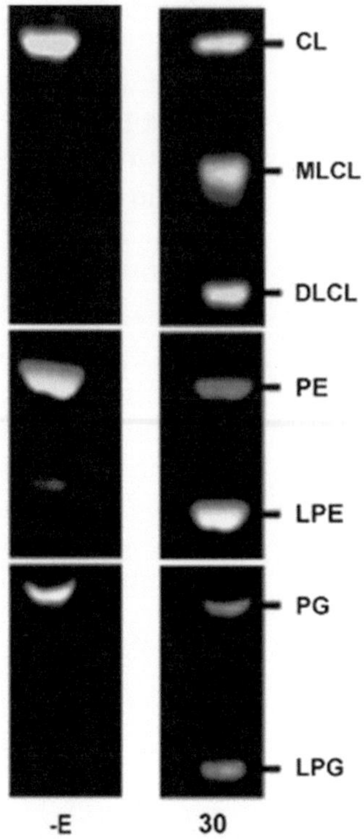

Fig. 4 Yeast phospholipase assays using fluorescent-labeled substrates. E, reaction without enzyme; 30, reaction product after 30 min of lipase assay

100 μl for 15 min at 30 °C. The reaction is stopped after the incubation time, and the lipids are resolved on a TLC plate using chloroform:methanol:25% ammonia (65:25:5, v/v). This separates the phospholipids and free fatty acids (FFA) simultaneously (Fig. 4).

1.2.8 Mass Spectrometric Tracking of PLs

Phospholipid tracking has come a long way from the traditional paper chromatographic methods of the early 1960s [68] to the most sophisticated mass spectrometric resolution wherein even the lipid species attached to the glycerol backbone can be easily identified as a function of their *m/z* ratios. Mass spectrometric methods have been successfully employed since the early 2000s in the analysis of lipids [69–71], with the biggest advantage being the ability to detect individual lipids from complex mixtures. With time, more specific methods for the detection of polar lipids have come in use. One such method has been described by Guan and Wenk [72]. It is a method based on untargeted profiling, and quantification using tandem mass spectrometry and multiple reaction monitoring, all of these using a single extract. While Q-Tof systems are used for ESI-based mass spectrometry, Tandem mass spectrometry for characterization/identification and quantification of lipid molecular species is done using a Q-Trap machine by precursor ion scanning (PREIS) and multiple reaction monitoring, respectively. Extensive studies on PC and PE molecular species profiles have also been reported by Anaokar et al. [73], where mass spectrometry has been used.

Based on recent reports that call for the detection of protein–lipid interactions that could be of physiological relevance, several new lipidomic approaches that use chromatography and mass spectrometry have been developed. Functional lipidomics that provides an integrative strategy of identifying the lipid, its target molecule, and their interactions are gaining prominence in the drug industry. This strategy has led to the discovery of unusual lipids like PC containing PUFAs and unique PIs [74, 75]. Another new group of PLs is that of oxidized PLs that are well recognized as biomarkers for oxidative stress. Mass Spectrometry has provided solutions for the detection of these lipids as well. Further, development of improved technologies like ion mobility, newer software for data analyses and imaging of the oxidised PLs can help in advancing our understanding of their physiology [76]. Oxidized phospholipids (OxPLs) have also been studied extensively using Q-ToF and Q-trap mass spectrometry by Aoyagi et al. [77].

2 Conclusion

Phospholipids thus play a very major and significant role in the cell, as a critical component of the cell membranes and as signal molecules. Recent evidence is proving their indispensable role in cell metabolism and pathology, making it necessary to study them in greater detail, in terms of quantification, localization, and upstream and downstream processes. Phospholipids are sure to emerge as the biomarkers for many of the human pathologies in the years to come, and the budding yeast *Saccharomyces cerevisiae* is a perfect model to simulate and study these conditions.

References

1. Fahy E, Subramaniam S, Brown HA, Glass CK, Merrill AH Jr, Murphy RC, Raetz CR, Russell DW, Seyama Y, Shaw W, Shimizu T, Spener F, van Meer G, VanNieuwenhze MS, White SH, Witztum JL, Dennis EA (2005) A comprehensive classification system for lipids. J Lipid Res 46:839–861
2. Bittman R (2013) Glycerolipids: chemistry. In: Roberts GCK (ed) Encyclopedia of biophysics. Springer, Berlin
3. Munnik T (2001) Phosphatidic acid: an emerging plant lipid second messenger. Trends Plant Sci 6(5):227–233
4. Taylor CL (2003) Phosphatidylserine and cognitive dysfunction and dementia (qualified health claim: final decision letter). Center for Food Safety and Applied Nutrition, USFDA. Retrieved 23 Aug 2014
5. Lands WE (1965) Lipid metabolism. Annu Rev Biochem 34:313–346
6. Wang B, Tontonoz P (2019) Phospholipid remodelling in physiology and disease. Annu Rev Physiol 81(1):165–188
7. Schneiter R, Kohlwein SD (1997) Organelle structure, function, and inheritance in yeast: a role for fatty acid synthesis? Cell 88:431–434

8. Raychaudhuri S, Young BP, Espenshade PJ, Loewen C Jr (2012) Regulation of lipid metabolism: a tale of two yeasts. Curr Opin Cell Biol Aug 24(4):502–508
9. Klug L, Daum G (2014) Yeast lipid metabolism at a glance. FEMS Yeast Res 14:369–388
10. Singh P (2016) Budding yeast: an ideal backdrop for *in vivo* lipid biochemistry. Front Cell Dev Biol 4:156. https://doi.org/10.3389/fcell.2016.00156
11. Claypool SM, Koehler CM (2012) The complexity of cardiolipin in health and disease. Trends Biochem Sci 37:32–41
12. Schuiki I, Daum G (2009) Phosphatidylserine decarboxylases, key enzymes of lipid metabolism. IUBMB Life 61:151–162
13. Lands WE (1960) Metabolism of glycerolipids. 2. The enzymatic acylation of lysolecithin. J Biol Chem 235:2233–2237
14. Wagner S, Paltauf F (1994) Generation of glycerophospholipid molecular species in the yeast *Saccharomyces cerevisiae*. Fatty acid pattern of phospholipid classes and selective acyl turnover at sn-1 and sn-2 positions. Yeast 10 (11):1429–1437
15. Fido M, Wagner S, Mayr H, Kohlwein SD, Paltauf F (1996) NATO ASI series. In: Op den Kamp JAF (ed) Molecular dynamics of biomembranes, vol H 96. Springer, Heidelberg, pp 315–326
16. Daum G (1985) Lipids of mitochondria. Biochim Biophys Acta 822:1–42
17. Yeagle PL (2016) Lipid protein interactions in membranes. In: Yeagle PL (ed) The membrane of cells, 3rd edn. Academic, New York
18. Chang S-C, Heacock PN, Clancey CJ, Dowhan W (1998a) The PEL1 gene (renamed PGS1) encodes the phosphatidylglycerophosphate synthase of *Saccharomyces cerevisiae*. J Biol Chem 273:9829–9836
19. Osman C, Haag M, Wieland FT, Brugger B, Langer T (2010) A mitochondrial phosphatase required for cardiolipin biosynthesis: the PGP phosphatase Gep4. EMBO J 29:1976–1987
20. Chang S-C, Heacock PN, Mileykovskaya E, Voelker DR, Dowhan W (1998b) Isolation and characterization of the gene (*CLS1*) encoding cardiolipin synthase in *Saccharomyces cerevisiae*. J Biol Chem 273:14933–14941
21. Tuller G, Hrastnik C, Achleitner G, Schiefthaler U, Klein F, Daum G (1998) YDL142c encodes cardiolipin synthase (Cls1p) and is non-essential for aerobic growth of *Saccharomyces cerevisiae*. FEBS Lett 421:15–18
22. Pangborn MC (1947) The composition of cardiolipin. J Biol Chem 168:351–361
23. Lecocq J, Ballou CE (1964) On the structure of cardiolipin. Biochemistry 3:976–980
24. Dimmer KS, Scorrano L (2006) (De)constructing mitochondria: what for? Physiology (Bethesda, Md) 21:233–241
25. Joshi AS, Thompson MN, Fei N, Huttemann M, Greenberg ML (2012) Cardiolipin and mitochondrial phosphatidylethanolamine have overlapping functions in mitochondrial fusion in *Saccharomyces cerevisiae*. J Biol Chem 287:17589–17597
26. Beyer K, Klingenberg M (1985) ADP/ATP carrier protein from beef heart mitochondria has high amounts of tightly bound cardiolipin, as revealed by 31P nuclear magnetic resonance. Biochemistry 24(15):3821–3826
27. Lange C, Nett JH, Trumpower BL, Hunte C (2001) Specific roles of protein-phospholipid interactions in the yeast cytochrome bc1 complex structure. EMBO J 20:6591–6600
28. Jiang F, Ryan MT, Schlame M, Zhao M, Gu Z, Klingenberg M, Pfanner N, Greenberg ML (2000) Absence of cardiolipin in the crd1 null mutant results in decreased mitochondrial membrane potential and reduced mitochondrial function. J Biol Chem 275:22387–22394
29. Gohil VM, Thompson MN, Greenberg ML (2005) Synthetic lethal interaction of the mitochondrial phosphatidylethanolamine and cardiolipin biosynthetic pathways in *Saccharomyces cerevisiae*. J Biol Chem 280:35410–35416
30. Schlame M, Rustow B (1990) Lysocardiolipin formation and reacylation in isolated rat liver mitochondria. Biochem J 272:589–595
31. Xu Y, Kelley RI, Blanck TJ, Schlame M (2003) Remodeling of cardiolipin by phospholipid transacylation. J Biol Chem 278:51380–51385
32. Schlame M, Kelley RI, Feigenbaum A, Towbin JA, Heerdt PM, Schieble T, Wanders RJ, DiMauro S, Blanck TJ (2003) Phospholipid abnormalities in children with Barth syndrome. J Am Coll Cardiol 42:1994–1999
33. Vreken P, Valianpour F, Nijtmans LG, Grivell LA, Plecko B, Wanders RJ, Barth PG (2000) Defective remodeling of cardiolipin and phosphatidylglycerol in Barth syndrome. Biochem-Biophys Res Commun 279:378–382. https://doi.org/10.1006/bbrc.2000.3952
34. Beranek A, Rechberger G, Knauer H, Wolinski H, Kohlwein SD, Leber R (2009) Identification of a cardiolipin-specific phospholipase encoded by the gene *CLD1* (YGR110W) in yeast. J Biol Chem 284(17):11572–11578
35. Rijken PJ, Houtkooper RH, Akbari H, Brouwers JF, Koorengevel MC, de Kruijff B,

Frentzen M, Vaz FM, de Kroon AI (2009) Cardiolipin molecular species with shorter acyl chains accumulate in *Saccharomyces cerevisiae* mutants lacking the acyl coenzyme A-binding protein Acb1p: new insights into acyl chain remodeling of cardiolipin. J Biol Chem 284:27609–27619

36. Ye C, Lou W, Li Y, Chatzispyrou IA, Huttemann M, Lee I, Houtkooper RH, Vaz FM, Chen S, Greenberg ML (2014) Deletion of the cardiolipin-specific phospholipase Cld1 rescues growth and life span defects in the tafazzin mutant: implications for Barth syndrome. J Biol Chem 289:3114–3125
37. Pokorna L, Cermakova P, Horvath A, Baile MG, Claypool SM, Griac P, Malinsky J, Balazova M (2015) Specific degradation of phosphatidylglycerol is necessary for proper mitochondrial morphology and function. Biochim Biophys Acta 1857:34–45
38. Nie J, Hao X, Chen D, Han X, Chang Z, Shi Y (2010) A novel function of the human CLS1 in phosphatidylglycerol synthesis and remodeling. Biochim Biophys Acta 1801:438–445
39. Lee SJ, Zhang J, Choi AM, Kim HP (2013) Mitochondrial dysfunction induces formation of lipid droplets as a generalized response to stress. Oxidative Med Cell Longev 2013:327167
40. Zweytick D, Athenstaedt K, Daum G (2000) Intracellular lipid particles of eukaryotic cells. Biochim Biophys Acta 1469:101–120
41. Farese RV Jr, Walther TC (2009) Lipid droplets finally get a little R-E-S-P-E-C-T. Cell 139:855–860
42. Murphy DJ (2001) The biogenesis and functions of lipid bodies in animals, plants and microorganisms. Prog Lipid Res 40:325–438
43. Tauchi-Sato K, Ozeki S, Houjou T, Taguchi R, Fujimoto T (2002) The surface of lipid droplets is a phospholipid monolayer with a unique fatty acid composition. J Biol Chem 277:44507–44512
44. Horvath SE, Wagner A, Steyrer E, Daum G (2011) Metabolic link between phosphatidylethanolamine and triacylglycerol metabolism in the yeast *Saccharomyces cerevisiae*. Biochim Biophys Acta 1811:1030–1037
45. Novikoff AB, Novikoff PM, Rosen OM, Rubin CS (1980) Organelle relationships in cultured 3T3-L1 preadipocytes. J Cell Biol 87:180–196
46. Pu J, Ha CW, Zhang S, Jung JP, Huh WK, Liu P (2011) Interactomic study on interaction between lipid droplets and mitochondria. Protein Cell 2:487–496
47. Shaw CS, Jones DA, Wagenmakers AJ (2008) Network distribution of mitochondria and lipid droplets in human muscle fibres. Histochem Cell Biol 129:65–72
48. Petit JM, Maftah A, Ratinaud MH, Julien R (1992) 10-N nonyl acridine orange interacts with cardiolipin and allows the quantification of this phospholipid in isolated mitochondria. Eur J Biochem 209:267–273
49. Gallet PF, Maftah A, Petit JM, Denis-Gay M, Julien R (1995) Direct cardiolipin assay in yeast using the red fluorescence emission of 10-N-nonyl acridine orange. Eur J Biochem 228:113–119
50. Jacobson J, Duchen MR, Heales SJ (2002) Intracellular distribution of the fluorescent dye nonyl acridine orange responds to the mitochondrial membrane potential: implications for assays of cardiolipin and mitochondrial mass. J Neurochem 82:224–233
51. Mileykovskaya E, Dowhan W, Birke RL, Zheng D, Lutterodt L, Haines TH (2001) Cardiolipin binds nonyl acridine orange by aggregating the dye at exposed hydrophobic domains on bilayer surfaces. FEBS Lett 507:187–190
52. Morita SY, Terada T (2015) Enzymatic measurement of phosphatidylglycerol and cardiolipin in cultured cells and mitochondria. Sci Rep 5:11737
53. Boumann HA, Gubbens J, Koorengevel MC, Oh CS, Martin CE, Heck AJR, Patton-Vogt J, Henry SA, de Kruijff B, de Kroon AIPM (2006) Depletion of phosphatidylcholine in yeast induces shortening and increased saturation of the lipid acyl chains: evidence for regulation of intrinsic membrane curvature in a eukaryote. Mol Biol Cell 17(2):1006–1017
54. Flis VV, Fankl A, Ramprecht C, Zellnig G, Leitner E, Hermetter A, Daum G (2015) Phosphatidylcholine supply to peroxisomes of the yeast *Saccharomyces cerevisiae*. PLoS One 10 (8):e0135084
55. Athenstaedt K, Daum G (2005) Tgl4p and Tgl5p, two triacylglycerol lipases of the yeast *Saccharomyces cerevisiae* are localized to lipid particles. J Biol Chem 280(45):37301–37309
56. De Kroon AI (2007) Metabolism of phosphatidylcholine and its implications for lipid acyl chain composition in *Saccharomyces cerevisiae*. Biochim Biophys Acta 771(3):343–352
57. Rockenfeller P, Koska M, Pietrocola F, Minois N, Knittelfelder O, Sica V, Franz J, Carmona-Gutierrez D, Kroemer G, Madeo F (2015) Phosphatidylethanolamine positively regulates autophagy and longevity. Cell Death Differ 22:499–508
58. Broekhuyse RM (1968) Phospholipids in tissues of the eye. I. Isolation, characterization

and quantitative analysis by two-dimensional thin-layer chromatography of diacyl and vinyl-ether phospholipids. Biochim Biophys Acta 152(2):307–315

59. Leventis PA, Grinstein S (2010) The distribution and function of phosphatidylserine in cellular membranes. Annu Rev Biophys 39:407–427
60. Casilly CD, Reynolds TB (2018) PS, it's complicated: the roles of phosphatidylserine and phosphatidylethanolamine in the pathogenesis of *Candida albicans* and other microbial pathogens. J Fungi (Basel) 4(1):E28. https://doi.org/10.3390/jof4010028
61. Kay JG, Grinstein S (2011) Sensing phosphatidylserine in cellular membranes. Sensors (Basel) 11(2):1744–1755
62. Nikawa J, Yamashita S (1997) Phosphatidylinositol synthase from yeast. Biochim Biophys Acta 1348(1–2):173–178
63. De Camilli P, Emr SD, McPherson PS, Novick P (1996) Phosphoinositides as regulators in membrane traffic. Science 271 (5255):1533–1539
64. Strahl T, Hama H, DeWald DB, Thomer J (2005) Yeast phosphatidylinositol 4-kinase, Pik1, has essential roles at the Golgi and in the nucleus. J Cell Biol 171(6):967–979
65. Wera S, Bergsma JCT, Thevelein JM (2001) Phosphoinositides in yeast: genetically tractable signalling. FEMS Yeast Res 1(1):9–13
66. Idevall-Hagren O, De Camilli P (2015) Detection and manipulation of phosphoinositides. Biochim Biophys Acta 1851(6):736–745
67. Bligh EG, Dyer WJ (1959) A rapid method of total lipid extraction and purification. Can J Biochem Physiol 37:911–917
68. Letters R (1966) Phospholipids of yeast II. Extraction, isolation and characterisation of yeast phospholipids. Biochim Biophys Acta 116(3):489–499
69. Faergeman NJ, Feddersen S, Christiansen JK, Larsen MK, Schneiter R, Ungermann C, Mutenda K, Roepstorff P, Knudsen J (2004) Acyl-CoA-binding protein, Acb1p, is required for normal vacuole function and ceramide synthesis in *Saccharomyces cerevisiae*. Biochem J 380(Pt 3):907–918
70. Forrester JS, Milne SB, Ivanova PT, Brown HA (2004) Computational lipidomics: a multiplexed analysis of dynamic changes in membrane lipid composition during signal transduction. Mol Pharmacol 65(4):813–821
71. Wenk MR (2005) The emerging field of lipidomics. Nat Rev Drug Discov 4(7):594–610
72. Guan XL, Wenk MR (2006) Mass spectrometry-based profiling of phospholipids and sphingolipids in extracts from *Saccharomyces cerevisiae*. Yeast 23(6):465–477
73. Anaokar S, Kodali R, Jonik B, Renne MF, Brouwers JFHM, Lager I, deKroon AIPM, Patton-Vogt I (2019) The glycerophosphocholine acyltransferase Gpc1 is part of a phosphatidylcholine (PC)-remodeling pathway that alters PC species in yeast. J Biol Chem 294:1189–1201
74. Koeberle A, Shindou H, Koeberle SC, Laufer SA, Shimizu T, Werz O (2013) Arachidonoyl-phosphatidylcholine oscillates during the cell cycle and counteracts proliferation by suppressing Akt membrane binding. Proc Natl Acad Sci USA 110(7):2546–2551
75. Koeberle A, Pergola C, Shindou H, Koeberle SC, Shimizu T, Laufer SA, Werz O (2015) Role of p38 mitogen-activated protein kinase in linking stearoyl-CoA desaturase-1 activity with endoplasmic reticulum homeostasis. FASEB J 29(6):2439–2449
76. Spickett CM, Pitt AR (2015) Oxidative lipidomics coming of age: advances in analysis of oxidized phospholipids in physiology and pathology. Antioxid Redox Signal 22 (18):1646–1666
77. Aoyagi R, Ikeda K, Isobe Y, Arita M (2017) Comprehensive analyses of oxidized phospholipids using a measured MS/MS spectra library. J Lipid Res 58(11):2229–2237

Chapter 5

What Can MS, NMR, and TLC Tell Us About the Composition of Lipid Membranes?

Kathrin M. Engel, Yulia Popkova, Jenny Leopold, and Jürgen Schiller

Abstract

Membranes protect the cells against the outer environment but are also very important to enable directed transport inside and outside the cell. In addition to selected membrane proteins and cholesterol, the membrane consists particularly of phospholipids, the contribution of which may be extremely different depending on the respective cell type. Methods for lipid analysis are diverse. Here, we will focus on lipid analysis by nuclear magnetic resonance (NMR) spectroscopy, thin-layer chromatography (TLC), and two different methods of mass spectrometry (MS), namely, matrix-assisted laser desorption/ionization (MALDI) and electrospray ionization (ESI). The advantages and disadvantages of these methods will be discussed based on human and boar spermatozoa lipid extracts, and the most relevant pitfalls in using these techniques will be outlined.

Keywords Phospholipids, MALDI-TOF MS, ESI MS, TLC, NMR spectroscopy, Lipid extracts, Spermatozoa

1 Introduction

All cells are separated against the outer environment by their lipid membrane. Additionally, intracellular membrane systems separate the different cellular compartments from each other. So different the individual cells can be, so different is the composition of their membranes. Cellular lipid composition is affected by the animal species, the surrounding temperature, the nutrition habits, the physiological function of the cell, and many other parameters [1].

1.1 Common Cellular Lipids

Biological membranes are assemblies of varying moieties of proteins, cholesterol, and (glycero)phospholipids [2]. The basic structure of the phospholipids (PLs) is derived from diacylglycerol, i.e., a glycerol backbone where two fatty acids are esterified with the free hydroxyl groups in *sn*-1 and *sn*-2 positions [3]. Sometimes there are also alkyl or alkenyl residues, i.e., the common ester bonds are replaced by ether linkages [4]. In most cases, there is a saturated

Rajendra Prasad and Ashutosh Singh (eds.), *Analysis of Membrane Lipids*, Springer Protocols Handbooks, https://doi.org/10.1007/978-1-0716-0631-5_5,

residue in the *sn*-1 position and an unsaturated one in the *sn*-2 position. If the *sn*-3 position is additionally esterified with phosphoric acid (H_3PO_4), the resulting product is called "phosphatidic acid" (PA). PA possesses one residual free hydroxyl group which may be esterified with compounds such as choline, ethanolamine, inositol, serine, or glycerol to yield phosphatidylcholine (PC), phosphatidylethanolamine (PE), phosphatidylinositol (PI), phosphatidylserine (PS), and phosphatidylglycerol (PG), respectively. In addition to these products there are often also products derived from sphingosine, particularly sphingomyelin (SM). Cardiolipins (CLs) are PL comprised of two PA molecules that are connected via a glycerol linkage. CL is particularly located in the mitochondrial membranes. In dependence on the respective organism, there may be tremendous qualitative and quantitative differences, but according to the best of our knowledge PA, PC, PE, PI, PS, PG, SM, and CL are the most relevant PL species in mammals [5].

Usually, the first step of lipid analysis is the extraction of lipids from the biological system of interest by organic solvents [6]—often chloroform and methanol mixtures: methods using methanol and chloroform have been described already in 1959 by Bligh and Dyer [7]. However, alternatives also have been developed and compared to these established extraction methods [8]. For instance, the use of methyl *tert*-butyl ether (MTBE) is gaining increasing popularity [9] as MTBE has a lower density than water (making the organic phase the upper layer in contrast to chloroform-based methods). This makes MTBE extraction more suitable for high-throughput applications and automation. Additionally, the risk of getting impurities from the aqueous into the organic layer is reduced under these conditions.

Sample extraction has two advantages: (1) lipids can be separated in a single step from more polar constituents of biological membranes such as salts, sugars, and proteins because these are insoluble in organic solvents and (2) extraction enables the convenient enrichment of lipids because the applied organic solvents are volatile and can be easily evaporated at which the lipids are concentrated. It must be explicitly stated that the method used for lipid extraction has a considerable impact on the subsequent recovery of lipids because the ability of every organic solvent to extract different lipid classes varies, and thus, quantitative extraction of a special lipid class may be hampered. This is a particular problem if both very polar and very apolar lipids are of interest in just one sample [10]. Therefore, one should always consider changing the extraction method (more apolar solvent systems for apolar lipids and vice versa) if one particular lipid class, which should be present in the material of interest, is absent or present in only small amounts in the extracted lipid fraction.

1.2 Methods of Lipid Analysis

There are many potential methods of lipid analysis. Nowadays, mass spectrometry (MS)—either as a shotgun approach (i.e., without separation of the lipid mixture into the individual lipid classes) or combined with liquid chromatography—is the method of choice [11, 12]. Here, the advantages of thin-layer chromatography (TLC) [13], high-resolution nuclear magnetic resonance (NMR) [14], and two different mass spectrometric methods, namely, matrix-assisted laser desorption and ionization (MALDI) [15] and electrospray ionization (ESI) MS [16] will be discussed. The advantages and disadvantages of all relevant methods are summarized in Table 1.

1.2.1 NMR Spectroscopy

All PL basically contain five NMR-detectable nuclei: ^{17}O exhibits a very low sensitivity and represents a "quadrupole" nucleus which is even worse than the low natural abundance of ^{17}O (0.038%). The resulting significant broadening of resonances makes ^{17}O not applicable for compositional analyses. Principally, ^{15}N and ^{13}C are "good" NMR nuclei ($I = ½$ spin) but both are only present at very small natural abundances (0.37% and 1.1%, respectively) which would result in poor signal-to-noise ratios (S/N) and require, thus, long lasting measurements and/or high sample concentrations [17]. However, only small amounts of material are available if biological samples are of interest. Compared to these disadvantages, ^{1}H and ^{31}P are the optimum NMR nuclei in PL research. ^{1}H NMR spectroscopy is suitable to determine the content of saturated, monounsaturated, and polyunsaturated fatty acids [18]. However, ^{1}H NMR is not feasible to determine the individual (saturated) fatty acids, i.e., the discrimination between palmitoyl and stearoyl residues (which just differ in a bismethylene group) is not possible. Furthermore, ^{1}H NMR spectroscopy enables the quantification of triacylgylcerols (TAG) in complex lipid mixtures because TAG yields a characteristic pattern of resonances [19].

The other spectroscopic method, ^{31}P NMR, is able to provide the relative moieties of the different PL classes in a mixture very easily [20]. This approach allows the identification of virtually all PL (and the lysophospholipids derived thereof [21]) in even very complex mixtures and the simultaneous quantification by the integration of the individual, separated ^{31}P resonances. For this, only a single internal standard of known concentration is necessary. Although the concentrations of PL can be easily determined by this approach, ^{31}P NMR is not suitable to evaluate the fatty acyl compositions of the different PL classes: changes in the fatty acyl compositions change the chemical shift of the corresponding ^{31}P resonance only to a very small extent and are—in practice—not reliable to determine the saturation degree of the PL. In contrast, the differentiation between diacyl, alkyl-acyl, and alkenyl-acyl PL is possible.

The reader should explicitly note that the suppression of PL aggregation in NMR spectroscopy is necessary because any PL aggregation (as in water) results in severe line-broadening [22]

Table 1
Experimental approaches frequently used to identify the different lipid classes as well as the related fatty acyl compositions

Technique	Applicable to	Advantages	Disadvantages
Mass spectrometry (MS)			
MALDI	All (polar and apolar) lipid classes	Simple, fast, and sensitive; sample impurities (e.g., salts, detergents) are tolerated; may be easily combined with TLC (desorption technique)	Ion suppression when analyzing crude mixtures; limited quantitative data analysis due to irregularities of the crystallization process
ESI	Principally all lipid classes; apolar lipids are more difficult to analyze than polar compounds	Can be easily coupled to HPLC and TLC, therefore, suitable for the screening of complex mixtures; most frequently used in lipid analysis	Ion suppression when analyzing crude mixtures; individual lipid classes are detected with different sensitivities; sensitive to contaminants (salts, detergents)
Nuclear magnetic resonance (NMR) spectroscopy			
^{31}P NMR	Phospholipids (at least without derivatization)	Nondestructive; direct, absolute quantitative analysis of virtually all PL	Expensive equipment; relatively low sensitivity; high concentrations of detergent required
^{1}H NMR	All lipids	Nondestructive; more sensitive than ^{31}P NMR	Mixtures of lipids give crowded spectra; expensive equipment; highly purified solvents required (for diluted samples)
^{13}C NMR	All lipids	Nondestructive; enables structure determination (of purified compounds); provides excellent spectral resolution	Mixtures of lipids give crowded spectra; expensive equipment; poor sensitivity
Chromatography			
Thin-layer chromatography (TLC)	All lipid classes (solvent system must be adjusted according to the lipid mixture and the stationary phase)	Simple, fast, and inexpensive; well established method; can be easily implemented in all laboratories; densitometric analyses of spot intensities are possible	Provides only limited resolution

(continued)

Table 1
(continued)

Technique	Applicable to	Advantages	Disadvantages
High-performance liquid chromatography (HPLC)	All lipid and PL classes (after establishment of a suitable solvent system)	Differentiation of individual lipid classes (normal phase) and fatty acyl composition (reversed phase); many well established methods available	Relatively large amounts of solvents required; risk of memory effects
Gas chromatography (GC)	Works exclusively with volatile compounds; hydrolysis and derivatization of lipids is required prior to fatty acid analysis	Method of choice to determine the fatty acid composition of mixtures; excellent resolving power	Requires volatile compounds and/or derivatization to enhance volatility. Not applicable to PL but only to released fatty acids (e.g., by saponification)

Reprinted with permission and modification from Methods Mol Biol. 1609:107–122

which makes unambiguous peak assignments impossible. This can be either done by using different solvent mixtures (such as chloroform, methanol, and a tiny amount of water) or by applying a dedicated detergent [23]. Since sodium cholate forms very small micelles in water (with an aggregation number of just four), this compound is the detergent of choice to investigate complex PL mixtures.

1.2.2 Mass Spectrometry

Nowadays, there are many different methods of MS which differ particularly in (I) the method of ion generation (ion source) and (II) the mass analyzer which sorts the generated ions in dependence on their molecular weights (or more precise the m/z ratio). We will focus here on MALDI-TOF MS and ESI-IT MS because both methods are routinely used in our laboratory. There are major differences between both techniques which can be summarized as follows [24]:

1. A solid sample (co-crystals between the analyte and the matrix) is used in the case of MALDI-TOF MS, while a solution of the analyte (without the need of matrix addition) is used in the case of ESI MS.
2. ESI is much more affected by sample impurities such as the (unwanted) salt content of the sample. Already millimolar concentrations of salt lead to a complete signal suppression while MALDI mass spectra (even if with reduced spectral quality) can be still recorded from samples with physiological salt content (154 mM NaCl).

3. According to the lucky survivor model [25] nearly exclusively singly charged lipid ions are generated during the MALDI process. In contrast, ions with several charges are easily generated during ESI MS—particularly if compounds with several charged groups (polyelectrolytes) and/or higher molecular weights are analyzed.

Unfortunately, both ESI and MALDI are affected by ion suppression effects. This means that the individual analytes are not necessarily detected according to their concentration but much more according to their tendency to be ionized. For instance, the quaternary ammonia group of PC (as well as SM and LPC) possesses a permanent positive charge. Thus, PC species are most sensitively detected in the positive ion mode and may even completely suppress the detection of other less sensitively detectable PL [26]. Accordingly, PC is normally not detectable at all in the negative ion mode—or only with very low sensitivity [27]. This aspect will be comprehensively discussed later.

1.2.3 Thin-Layer Chromatography

TLC is an old but still powerful method for lipid separation [13]. TLC can be conveniently performed, is relatively inexpensive, and can be easily established in every laboratory—also without dedicated and sophisticated analytical equipment. The simple establishment is probably the most important advantage of TLC compared to LC. However, there are some additional advantages of TLC which were recently summarized [13]. The most obvious advantage of TLC is probably the complete exclusion of memory effects: while an LC column is used many times, a new stationary phase is used in every TLC experiment. This excludes the risk of detecting products from a previous experiment.

Normal-phase (NP) silica gel chromatography is traditionally used in TLC [28]. Under these conditions PL are separated primarily according to their different polar headgroups—not according to differences in the acyl groups. If the acyl compositions in addition to the headgroups are also of interest, two-dimensional TLC experiments with reversed phase (RP) [29] or silver ion impregnated plates are advisable [30]. The latter approach enables the separation of lipids in dependence on the number of double bonds to which Ag^{+} will bind. Recently, it has been shown that oxidation products of PL can be more readily separated by RP-TLC than NP-TLC [31] which emphasizes the power of this approach.

2 Materials

2.1 Equipment and Supplies

2.1.1 ^{1}H and ^{31}P NMR Spectroscopy

1. High-field NMR spectrometer (for instance from Bruker Biospin, Rheinstetten, Germany) (*see* **Note 1**).
2. Three channel NMR probe tunable to the frequency of ^{31}P (first channel) and ^{1}H decoupling in the second channel. ^{1}H decoupling minimizes the line-widths of the individual resonances and, thus, improves the achievable *S/N* ratio significantly. Finally, a "lock" (^{2}H) channel should be available to compensate for the drift of the magnetic field by fixing the ^{2}H signal of the deuterated solvent (normally D_2O). The "lock" is particularly important if long lasting measurements are performed. Finally, accurate temperature control should be possible to warrant that all spectra are acquired at the same temperature and to avoid potential changes due to environmental effects.
3. High-quality NMR sample tubes with a diameter of 5 mm (*see* **Note 2**).
4. Micropipettes for aqueous solutions (*see* **Note 3**) and glass (Hamilton) syringes for chlorinated hydrocarbons (normally used to dissolve PL).
5. Sonifier or ultrasound bath (preferentially with heating) to diminish the extent of PL aggregation.

2.1.2 ESI-IT MS

1. ESI-IT mass spectrometer (in our case, an Amazon SL® device, Bruker Daltonik GmbH, Bremen, Germany) equipped with a syringe pump and, in the optimum case, with an extractor (for instance from Advion, Ithaca, NY, USA) which enables the direct release of the different PL classes from the developed TLC plate (*see* also **Note 4**).
2. Micropipettes for aqueous solutions (*see* **Note 3**).
3. Glass (Hamilton) syringes of different sizes (for solvents such as $CHCl_3$).
4. Ultrasound bath to degas the solvents needed for the TLC extractor.

2.1.3 MALDI-TOF MS

1. MALDI-TOF mass spectrometer (for instance an Autoflex® device from Bruker Daltonik GmbH) equipped with a reflectron (enhances the ion flight path and in that way improves the resolution), "delayed extraction (DE) facility" and a laser emitting at 337 or 355 nm. The acquisition of positive and negative ion spectra should be possible (*see* also **Note 5**).
2. MALDI targets made from stainless steel or from aluminum (*see* **Note 6**).

3. Micropipettes (*see* **Note 3**).
4. Glass (Hamilton) syringes of different sizes.
5. Single-use glass vessels (available, for instance, from Knauer Wissenschaftliche Geräte GmbH, Berlin, Germany) for mixing the matrix solution and the lipid samples typically dissolved in chloroform.
6. Heat gun or common hair dryer (*see* **Note 7**).

2.1.4 Thin-Layer Chromatography

1. Normal-phase silica-coated glass or aluminum plates from MERCK KGaA, Darmstadt, Germany or any other company (*see* **Note 8**).
2. Solvent trough which fits to the size of the used TLC plates (normally 10 × 10 cm or 20 × 10 cm).
3. Nebulizer, evaporator, or dipping apparatus to cover the TLC plates with a suitable dye.
4. UV lamp and/or a hot plate to visualize the spots on the TLC plate.

2.2 Reagents and Biological Mixtures

1. Stock solutions of 1-palmitoyl-2-oleoyl-*sn*-PC (POPC), -PE (POPE), and -PG (POPG) in $CHCl_3$ (e.g., from Avanti Polar Lipids, Alabaster, AL, USA). These stock solutions should be diluted with chloroform to a concentration of 1 mg/ml which is a good starting point for all relevant methods (*see* **Note 9**).
2. 2,5-Dihydroxybenzoic acid (DHB) and 9-aminoacridine (9-AA) as MALDI matrices. DHB is dissolved in methanol to obtain a 0.5 M (about 77 mg/ml) solution, whereas a 10 mg/ml solution of 9-AA is prepared in isopropanol/acetonitrile (60:40, v/v) [32]. Matrices and solvents should be of the highest commercially available quality.
3. Solvents of the highest commercially available purity for ESI-IT MS, TLC and for the direct elution of lipid fractions from the TLC plate into the ESI machine.
4. Primuline ("Direct Yellow 59") for visualizing lipids on the TLC plates. For spraying, a solution of 0.05 mg/ml in acetone/water (80:20, v/v) is recommended. For dipping the plate, a 0.5 mg/ml primuline solution in acetone/water (80:20, v/v) should be used.
5. 50 mM 4-(2-hydroxyethyl)piperazine-1-ethanesulfonic acid (HEPES) buffer (pH 7.65) containing 200 mM sodium cholate (to suppress the aggregation of PL which typically occurs in water in the absence of detergents) and 5 mM ethylenediaminetetraacetic acid (EDTA) (to reduce the line-broadening effects of paramagnetic ions such as ferrous or cupric ions which may be bound to negatively charged PL). This solution is best prepared in D_2O to provide a sufficient field frequency

lock. The slight differences in the pK values of H_2O and D_2O are negligible because the pH is adjusted in the final step of the buffer preparation (*see* **Note 10**).

6. A small amount of the biological fluid, the cells, or the tissue of interest. Lipids can be easily extracted by the following procedure: to increase the extraction yield samples can be homogenized using a tissue homogenizer, for instance, Precellys® 24 (VWR, Radnor, Pennsylvania, USA). Different ball mill tubes and settings can be used for this approach to optimize the lipid yield. Afterwards, the biological material is treated with the about 20-fold amount (by weight) of Bligh and Dyer solvent [7] mixture (chloroform:methanol:water = 1:1:1) or any other suitable solvent system and the resulting turbid mixture vortexed for a few minutes. Subsequently, centrifuge the sample (10 min, 2000 rpm (about 800 g), RT) in order to improve the separation of the organic (bottom) and the aqueous phase (*see* also **Note 11**). Remove the organic (lower) chloroform phase by a Hamilton syringe and transfer it to another unused vial.
7. To concentrate the sample, the organic fraction can be evaporated to dryness either under a stream of nitrogen or by a vacuum concentrator. Afterwards the sample is redissolved in a small volume of the desired solvent.

3 Methods

3.1 Sample Processing

3.1.1 MS Characterization of Artificial Lipid Samples of Known Composition

1. The use of lipid standard stock solutions in the 0.1 mg/ml concentration range is recommended to check the performance of the used mass spectrometer—although MS is much more sensitive and would even detect analytes in the pg range. If large lipid amounts are available, dilute the lipid samples to about 0.1 mg/ml with chloroform. Mix one equivalent of the lipid standard solution with either one volume equivalent of DHB or 9-AA matrix solution. Vortex for good homogeneity.
2. Apply the sample/matrix mixtures to the MALDI target. Avoid touching the MALDI target with the pipette tip or the needle of the Hamilton syringe as this impairs homogeneous crystallization. Avoid potential contaminations of the samples by skin lipids (e.g., from your hands) by wearing gloves.
3. Evaporate the solvent as quickly as possible (either by applying vacuum or by a stream of warm air). Subsequently load the MALDI target into the mass spectrometer. Avoid long-term exposition of the target with the deposited samples to air as this might cause unwanted oxidation of unsaturated lipids (*see* **Note 12**).

4. ESI mass spectra can be recorded in the same way with the exception that the lipid solution is infused into the MS device by a syringe pump. However, the use of apolar solvents as $CHCl_3$ is discouraged; more polar solvents such as isopropanol or methanol should be used instead.

3.1.2 MS of Lipids from Spermatozoa Extracts

1. The obtained crude (human and boar) sperm lipid extract (vide supra, Subheading 5.2.2) may be directly used and diluted with the prepared matrix solution (MALDI) or a suitable solvent (ESI) in the desired volume ratio (*see* **Note 13**).
2. Use the same method of sample preparation as described in the context of the artificial lipids (Subheading 5.3.1.1).

3.1.3 Separation of Human and Boar Sperm Lipid Extracts by Normal-Phase One-Dimensional TLC

1. The obtained crude sperm lipid extract (vide supra, Subheading 5.2.2) may be directly used for TLC separation.
2. Apply the sperm extract and a standard PL mixture, which contains different PLs (known composition and concentration) onto the TLC plate. Automated sample application is highly recommended, e.g., by using a LINOMAT device (CAMAG, Berlin, Germany).
3. Develop the TLC plate using a suitable solvent system. For PL mixtures, the solvent system chloroform/ethanol/water/ triethylamine (30:35:7:35, v/v/v/v) gives a reasonable separation of lysolipids (LPC, LPE, lyso-CL), SM, PC, PS, PI, PE, and free fatty acids. Even the separation of one PL class into short- and long-chained or diacyl- and alkyl-acyl-linked PL can be achieved. However, the separation of PA/PS, PG/CL, and cholesterol/triacylglycerols is relatively poor using this solvent system.
4. Visualize the separated lipid spots by primuline staining. Primuline can either be applied by spraying or by dipping (by hand or automatically). Dry the plate carefully by a common hair dryer. Lipid spots are visible under UV light (366 nm) and can be marked by a pencil for subsequent extraction.
5. Separated lipids can be analyzed either by MALDI MS (as essentially described in [33]) or by direct elution and infusion into a mass spectrometric device, e.g., an ESI-IT mass spectrometer—as initially described by Luftmann [34].

3.1.4 Preparation of the ^{31}P NMR Samples

1. Use 100 μl of the original, concentrated spermatozoa lipid extract and evaporate the organic solvent either under a stream of nitrogen or by a vacuum concentrator to obtain a dried lipid film.
2. Resolubilize the lipid film in about 500 μl of the detergent solution described above (Subheading 5.2.2). Vortex the sample until the lipid film has been dissolved completely. If the

sample remains opaque, use ultrasound and/or slight heating to convert the obtained suspension into a clear "solution". This sample-detergent mixture can be directly used for ^{31}P NMR. However, it is advantageous to allow the sample to equilibrate for a few hours at room temperature (*see* also **Note 14**).

3.2 Recording MALDI-TOF Mass Spectra

1. Due to the large number of MALDI mass spectrometers, which are commercially available nowadays, it is impossible to describe the necessary experimental parameters in detail. Therefore, please consult the manual of your MS device for a suitable method file to start with. Nevertheless, you should start with "delayed extraction conditions" (DE) and make use of the reflectron of your MS to improve the spectral quality by resolution enhancement.
2. It is advisable to start with the analysis of a defined PL sample (standard) with known composition and concentration in order to check if the MS device is properly working and to verify whether all parameters are adequately set (*see* Subheading 5.3.1.1). This standard solution is also useful to check the mass accuracy and the quality of the applied mass calibration file. The applied laser fluency [which is adjusted by using a relative scale ranging from 0 to 100% (no attenuation at all, i.e., the laser irradiates the sample with its maximum power)] normally has the most pronounced effect on the spectral quality [35] and should be adjusted carefully.
3. Many MALDI users believe (in error) that the *S/N* ratio may be improved by increasing the laser fluence. However, this is not true: although peak intensities are enhanced at elevated laser intensities, the quality of the baseline is concomitantly deteriorated since more fragment ions are generated. Therefore, the laser intensity should be set as high as necessary but as low as possible.
4. Try to move the laser randomly (many MALDI mass spectrometers provide a "random walk" shot cycle) over the complete sample and average a large number of laser shots (100–2000 depending on the MS device) in order to average nonhomogeneous regions of the spot and to avoid distorted (subjective) spectra. However, averaging a larger number of individual laser shots does not improve the *S/N* ratio [36] because the level of unspecific chemical background noise embodies the limiting criterion.

3.3 Recording ESI-IT Mass Spectra

1. ESI-IT mass spectra can be acquired in a similar manner as the MALDI mass spectra (*see* **Note 15**). However, more polar solvents (e.g., isopropanol or methanol) should be used. Particularly the use of chloroform is discouraged because this

solvent may lead to the generation of unwanted (dichlorocarbene) adducts.

2. Due to considerable differences in the configuration of available ESI mass spectrometers, it is not possible to describe the setup of the experimental parameters in detail. Therefore, please consult the manual of your MS device.
3. In particular the flow rate (how many μl of sample are infused into the device per minute) and the temperature of the nebulizer gas as well as the gas pressure have to be carefully adjusted. It is also important to allow the mass analyzer to collect enough ions. This is sometimes challenging if fast flow rates are used.
4. Note that the purging of the tubes and fittings is very important to avoid the "carryover" of analytes from one sample to the next sample. Also always record the background of your solvent prior to the ESI analysis of your desired sample.

3.4 ^{31}P NMR Spectroscopy

1. Introduce the sample into the magnet of your NMR spectrometer. Make sure that the sample volume is centered between the coils (a sample volume of 500 μl is indicated in the case of 5 mm NMR tubes) (*see* **Note 16**).
2. Set the temperature to the desired value (310 K/37 °C is indicated for PL) and allow the sample to equilibrate. The required time depends on the geometry of your NMR spectrometer; about 20 min should be sufficient.
3. Find the lock (^{2}H) signal of the deuterated solvent and lock in. If your spectrometer has been carefully configured this is done automatically.
4. Tune and match the probe to ^{31}P (first channel) and ^{1}H (decoupling channel). On modern devices this will be done automatically.
5. Load a suitable dataset for ^{31}P with broadband decoupling.
6. Determine the 90° pulse for ^{31}P and the decoupling (^{1}H) pulse. Follow the instructions given in your NMR spectrometer handbook or in [17].
7. Determine the appropriate receiver gain and record a ^{31}P (^{1}H decoupled) spectrum. Ensure that the number of points (time domain) is sufficiently high to record the entire free induction decay (FID) and no part of the FID is truncated.
8. Apply Fourier transformation to convert the obtained raw data from intensity time to intensity frequency, i.e., to generate a regular "spectrum". Apply a line-broadening (LB) factor of less than 2 Hz because a higher value would make peaks with similar chemical shifts undistinguishable. Phase correct your spectrum and set the resonance of PC (normally the most intense peak at the right end of the spectrum) to −0.65 ppm.

9. Integrate all observed resonances subsequent to appropriate baseline correction (particularly important in the case of diluted samples with poor *S/N* ratio).

4 Results

A survey about the shape of the positive and negative ion MALDI-TOF mass spectra of different, selected PL (as well as their characteristic headgroups) in the presence of two different matrices is shown in Fig. 1 (reproduced from [37] with permission). The polarities of the individual measurements (positive or negative ion detection) are indicated directly at the spectra. POPC (**1a**), POPE (**1b**, **1e**), and POPG (**1c**, **1f**) were chosen as representative PL because they are abundant in many biological samples and differ characteristically in their charges [22]: POPC and POPE are zwitterionic PL, while POPG is negatively charged under physiological conditions (i.e., at pH 7.4). In (**1d**) and (**1g**), the positive and negative ion spectra of 1:1:1 (mol:mol:mol) mixtures of these PL are shown to illustrate the typical problem of mixture analysis by soft-ionization MS. Although exclusively the MALDI mass spectra are shown here, it is important to note that very similar spectra would be obtained if ESI MS instead of MALDI would have been used.

All the mass spectra differ significantly: in the presence of the (acidic) DHB matrix POPC yields two peaks at *m/z* 760.6 and 782.6 corresponding to the H^+ and the Na^+ adduct (**1a**), respectively. Due to the permanent positive charge of its headgroup group [38], PC is hardly detectable as negative ion [27, 39]. There are no further peaks, i.e., no fragmentation of POPC occurs in the positive ion mode. In contrast, POPE (**1b**) and POPG (**1c**) yield a characteristic fragment (*m/z* 577.5) corresponding to the loss of the headgroup [15]. It is also obvious that the POPE with a monoisotopic mass of 717.5 g/mol gives only rather small yields of the H^+ adduct (*m/z* 718.5) but high yields of the Na^+ adduct (*m/z* 740.5) and (to a minor extent) the Na^+ adduct subsequent to the exchange of one H^+ by one Na^+ (*m/z* 762.5) [39]. The latter adduct is caused by the exchangeable protons of the $-NH_3^+$ group of the PE and therefore not seen in the case of POPC.

Using 9-AA, a more alkaline matrix (pK $\approx$ 9.99) than DHB [39], POPE is detectable as negative ion at *m/z* 716.5 (**1e**). The intensity of this peak is very low if DHB is used. In contrast, POPG can be detected as negative ion in the presence of DHB (data not shown) as well as 9-AA (**1f**, *m/z* 747.5) because POPG is more acidic than the zwitterionic POPE.

The different detectabilities of the individual PL are of paramount interest if it comes to mixture analysis: in the positive ion mode (**1d**), the spectrum of the POPC/POPE/POPG equimolar

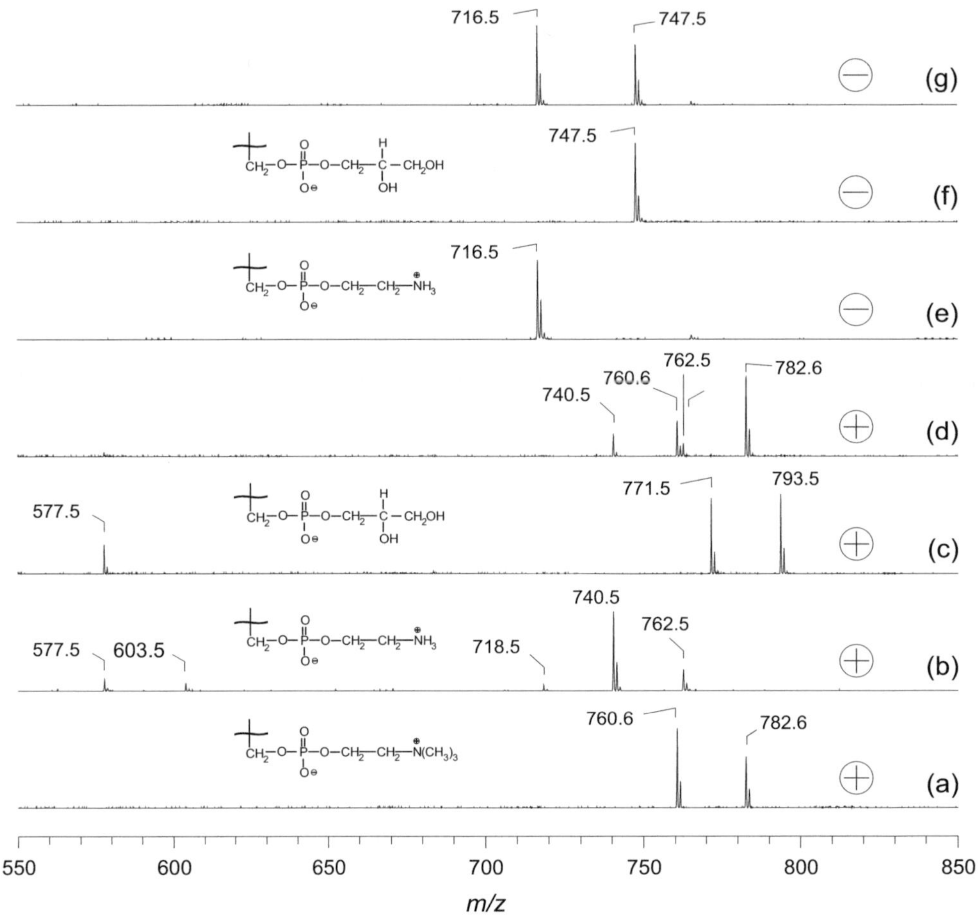

Fig. 1 Positive and negative ion MALDI-TOF mass spectra of 1-palmitoyl-2-oleoyl-*sn*-phosphatidylcholine (POPC; **a**), 1-palmitoyl-2-oleoyl-*sn*-phosphatidylethanolamine (POPE; **b**, **e**), 1-palmitoyl-2-oleoyl-*sn*-phosphatidylglycerol (POPG; **c**, **f**), and a 1:1:1 mixture of these three PLs (**d**, **g**). The polarities of the measurements are indicated directly at the individual spectra. Positive ion spectra were recorded with a 0.5 M solution of 2,5-dihydroxybenzoic acid (2,5-DHB) in methanol, whereas a 10 mg/ml solution of 9-aminoacridine (9-AA) in isopropanol/acetonitrile (60:40, v/v) was used as matrix for the negative ion spectra. PL sample solutions (0.1 mg/ml in chloroform) were diluted 1:1 (v/v) with the corresponding matrix and afterwards spotted onto the MALDI target. All peaks are marked according to their *m/z* ratios, and the structures of the headgroups of the relevant PL classes are also given. Reprinted with permission and modification from Methods Mol Biol. 1609:107–122

mixture is dominated by the POPC (*m/z* 760.6 and 782.6). POPE is detected only with much lower intensity (*m/z* 740.5 and 762.5). Due to its negative charge, POPG is not detectable under these conditions. Therefore, the presence of PC aggravates the detection of other PL species [26, 33]. In contrast, POPC is not detectable at all in the negative ion spectrum (**1g**) and only POPE and POPG are detectable in the PL mixture. Although this is indeed a very simple example, the reader should be aware that mixture analysis by

MALDI-TOF MS (as well as ESI MS) must be regarded with great caution—in particular if only spectra in one polarity mode (positive or negative) are acquired.

The problem of signal suppression is even more important if organic extracts of biological samples such as human and boar spermatozoa extracts with "unknown" compositions are analyzed (Fig. 2). The positive ion MALDI mass spectra were recorded with DHB while 9-AA served as matrix for the acquisition of the negative ion mode spectra. The polarities by which the spectra were acquired are indicated in the spectra. The most intense peaks are directly assigned to the corresponding PL. It is obvious that the positive ion mode spectra are dominated by PL with quaternary ammonia groups, i.e., PC and SM. At the first glance, this might be surprising because spermatozoa extracts are known to contain additional PL classes—particularly PE [40]. However, considering the ion suppression effect described above this is not surprising because both, SM and PC, possess a quaternary ammonia group which is always positively charged. This leads to the fact that the presence of PC and SM outshines all other PL classes in the positive ion mode. In many cases, this problem can be overcome by the acquisition of the negative ion mass spectra (either MALDI or ESI) because some PL classes are more readily detectable as negative ions. Unfortunately, in the case of spermatozoa, this is no suitable way out. The negative ion spectra are dominated by a very intense peak at *m/z* 795.6 which is assigned to the seminolipid, a sulfoglycolipid [41], the structure of which is shown in the figure. The sulfate residue in this lipid represents a strong electrolyte which is always negatively charged and, thus, suppresses all other PLs which would otherwise be detectable in the negative ion mode [42]. Therefore, the detection of PE in spermatozoa extracts is only possible by NMR or subsequent to the separation of the entire extract into the individual PL species by chromatographic means (Fig. 3). Separating the sperm lipid extract reveals that human spermatozoa are unequivocally composed of GPC and GPE (two spots due to the presence of diacyl- and alkyl-acyl-linked residues) (*see* **Note 17**), SM (two spots due to short- and long-chained SM), CL, and comparably small amounts of LPC, PS, PI as well as free fatty acids. It is also obvious that the distribution of the different PL species within the TLC spots is inhomogeneous [33]: the mass spectra recorded from the upper part of the TLC spots are dominated by alkyl-acyl-GPE species while alkenyl-acyl-GPE and diacyl-PE species are located in the bottom of the spots. Furthermore, the positive ion spectra are more complicated than the negative ion spectra which are characterized by a less complicated adduct pattern.

The relative amounts of each PL can be determined densitometrically [43]. However, it has to be taken into account that the primuline dye has a different affinity to the individual PL classes

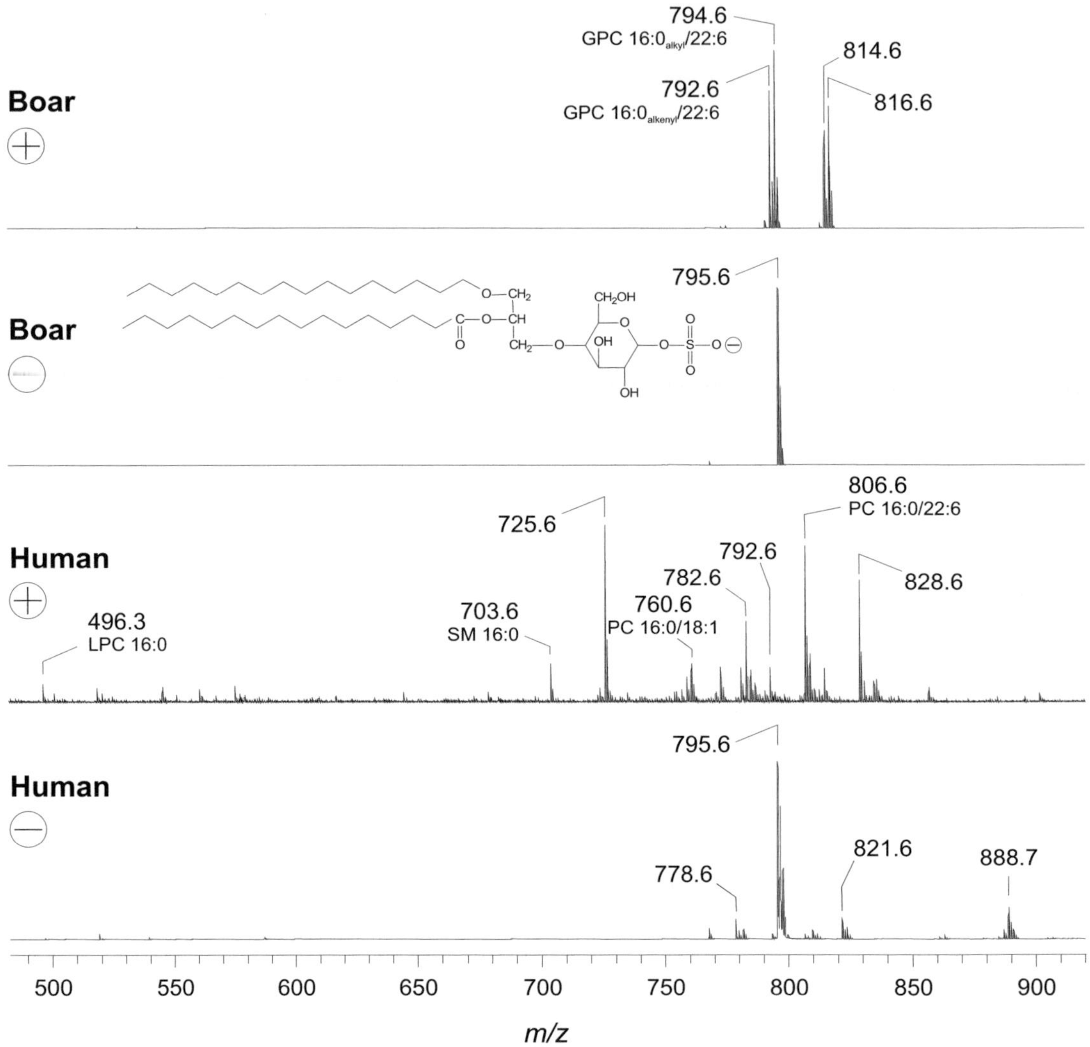

Fig. 2 Positive and negative ion MALDI-TOF mass spectra of an organic extract of boar and human spermatozoa. Polarities of the measurements and assignments of the most prominent peaks (only the proton adducts) are given at the relevant spectra. Positive ion spectra were recorded in the presence of DHB and negative ion spectra with 9-AA. Note the complete suppression of PE in the presence of PC in the positive ion mode spectra

and, therefore, different PLs are detected with different sensitivities. Thus, calibration curves have to be recorded for each PL if the PL concentrations on the TLC plate are to be determined quantitatively.

NMR has the advantage that relative quantitative information can be directly obtained from the integration of the spectra [17]. Absolute quantitative information can be obtained in the presence of just a single PL standard (which does not occur in the extracts of interest); this is a significant difference to MS where one

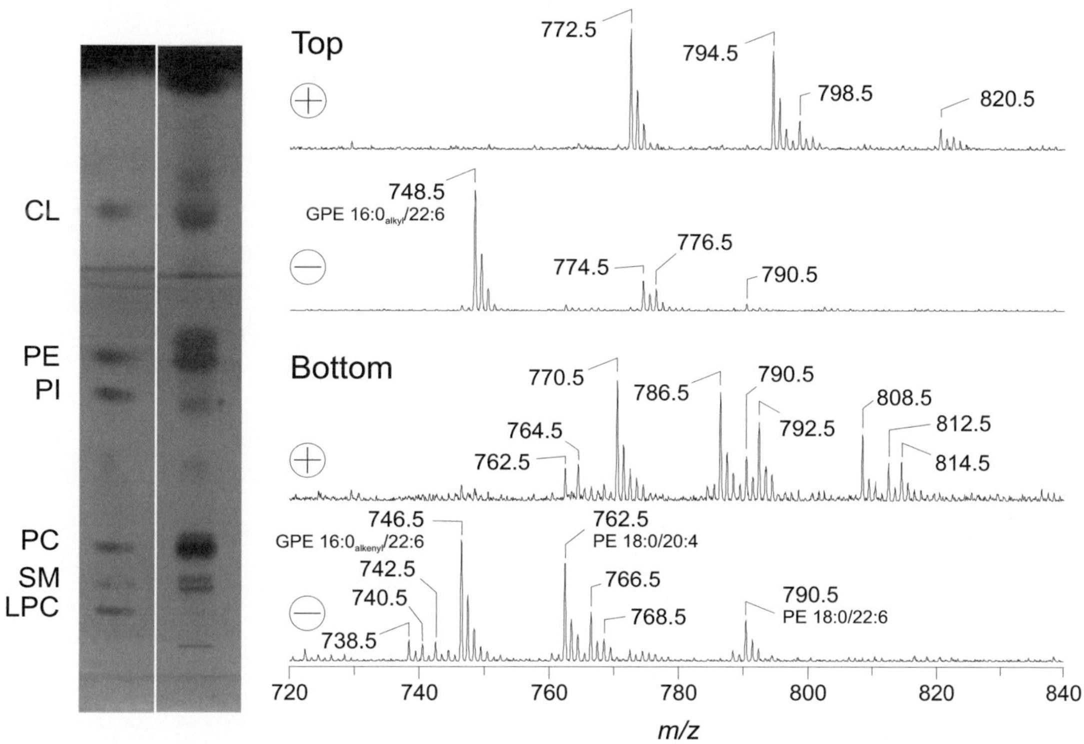

Fig. 3 Separation of a human sperm lipid extract by TLC. To assign the PL classes a standard lipid mixture (left panel) has been applied next to the sample (right panel). The TLC plate was developed with chloroform/ethanol/water/triethylamine (30:35:7:35, v/v/v/v) as the mobile phase. Lipids were afterwards visualized by dipping the plate into 0.5 mg/ml primuline (in acetone/water 80:20, v/v) and subsequent irradiation with UV light. Lipid spots were marked by a pencil. Identified PL fractions were analyzed by ESI-IT MS by direct elution from the TLC plate and infusion into the MS device. The PE fraction splits into two obvious subfractions which were analyzed separately (Top and Bottom). Polarities of the measurements and assignments of the most prominent peaks are given at the spectra

standard for each PL class is necessary. The ^{31}P NMR spectra of boar and human spermatozoa are compared in Fig. 4. Each PL is characterized by a dedicated chemical shift (given in ppm) which makes spectral separation of the PL quite simple [42]. The relative contributions of the different PL classes can be easily calculated into absolute amounts with the help of one internal standard which is present in a known concentration. It is also obvious that some PL resonances do not appear as singlets but are split into two resonances. This is caused by differences in the fatty acyl compositions (and/or the presence of alkenyl and alkyl species) which can be, however, only partially resolved by ^{31}P NMR [20].

In a nutshell, we have shown that the discussed analytical methods are useful to elucidate the lipid compositions of membranes. TLC can be used as a convenient and inexpensive method to monitor the presence of different PL classes in biological samples. When combined with MS detection, the apolar residues of the

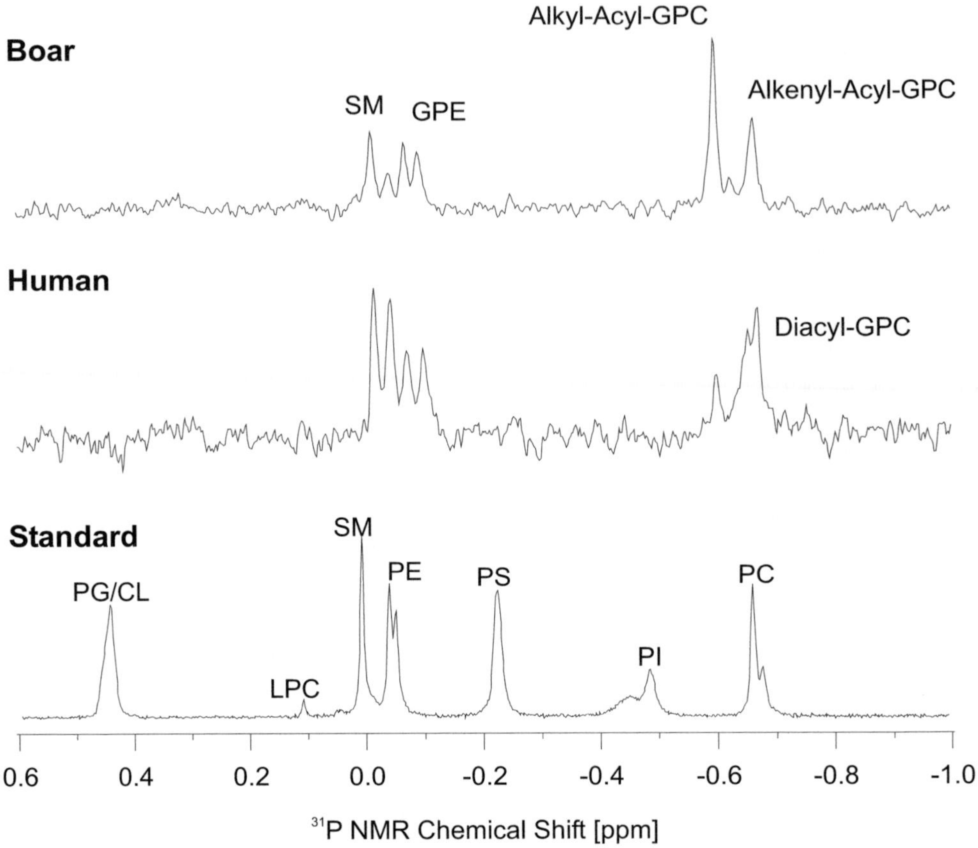

Fig. 4 High-resolution ^{31}P NMR spectra of organic extracts of boar and human spermatozoa. For comparative purposes, the spectrum of a PL mixture of known composition is shown at the bottom. All spectra were recorded in aqueous sodium cholate (pH 7.65) in order to suppress PL aggregation and, thus, to maximize resolution. All peaks are assigned to the corresponding PL class. Note that some PL species result in two resonances due to differences in the chain lengths and/or the presence of ether lipids

different PLs can also be resolved. However, one internal standard per PL class is needed to obtain quantitative information. This problem can be overcome by using high-resolution ^{31}P NMR, although NMR has the disadvantage that only poor information about the apolar residues can be achieved. Therefore, TLC, MS, and NMR complement each other perfectly.

5 Notes

1. Although there are recent reports that low-field NMR spectrometers (inexpensive, with an electric magnet) can be used to evaluate the PL compositions of complex mixtures [44], the use of high-field NMR spectrometers provides several advantages: (1) the spectral resolution is enhanced at higher field

strength. This means that PL with different acyl residues (i.e., with small chemical shift differences) can still be differentiated. Using a powerful magnet is particularly important for ^{1}H NMR spectra because the chemical shift range is relatively small and (2) the sensitivity of the NMR spectrometer increases with the magnetic flux density (B_0). This is important if biological, low-concentrated samples are of interest.

2. Nowadays, 5 mm NMR tubes are commonly used, and most NMR probes are designed for this sample size. NMR tubes which have this diameter (5 mm) at the top and a smaller diameter at the bottom of the tube are also commercially available and possess the advantage that the available sample may be concentrated since a reduced solvent volume is required. In contrast, 10 mm NMR sample tubes are nowadays rarely used because the advantage of a larger sample size (more sample within the coil volume) is accompanied by a reduced resolution because larger volumes decrease the quality of the "shim", i.e., the homogeneity of the magnetic field.
3. Compared to MS, the impact of impurities on the quality of NMR spectra is much less pronounced because NMR is by far less sensitive. Nevertheless, it is advisable to use exclusively glassware as soon as there is the need to use chlorinated hydrocarbons (CH). CH represents "aggressive" solvents that may leach impurities such as plasticizers or antioxidants (for instance, Irganox® or Irgafos®) from plastic material [45].
4. Basically, all available ESI mass spectrometers can be used. The indicated Amazon mass spectrometer is an inexpensive but low-resolution device. Higher resolution devices should be preferentially used because they provide (due to their high mass accuracy) the possibility to derive the chemical formula directly from the obtained *m/z* values. This often helps in the assignment of peaks. Additionally, "higher resolution" leads to "sharper" peaks, which goes along with improved sensitivity (the number of generated ions is distributed over a smaller peak area).
5. Basically, there is no need to use any specific MALDI mass spectrometer. However, it is recommended that a device with a reflector detector and with "delayed" extraction (sometimes also called "ion velocity focusing") capability is used in order to improve the resolution. Traditional "linear" devices (with a relatively short flight path) are characterized by a comparably poor resolution due to the reduced length of the flight tube and the enhanced peak widths. Optimum resolution is particularly important if smaller compounds are analyzed.
6. Some MS companies offer disposable MALDI "plastic" targets. These targets are designed for aqueous solutions. As soon as

CH is used, it has to be checked whether impurities are leached from the plastic material.

7. Most lipid samples will be extracted from the biological material by using chloroform/methanol mixtures. Therefore, lipids are commonly dissolved in chloroform—subsequent to the extraction. Unfortunately, many established matrix compounds (such as DHB) are not soluble in chloroform but require more polar solvents such as methanol. The necessity to apply different solvents (with different volatilities) leads to a major problem: inhomogeneous matrix/analyte co-crystals and, thus, low shot-to-shot reproducibility. This problem can be minimized by evaporating the solvent on the MALDI target as fast as possible, e.g., by using a common hairdryer.
8. According to our experience, the TLC plates from some manufacturers contain impurities which may lead to problems—particularly if subsequent MS analysis is planned. In that case, TLC plates with "MS grade" purity should be used [46]. These plates are characterized by a decreased silica layer thickness (100 vs. 200 μm of standard plates) and optimized purity of the silica gel.
9. It is always a good idea to check the performances of the used instruments by measuring selected compounds of known composition and concentration from time to time. PC and PG should be particularly sensitively detectable as positive and negative ions, respectively, while PE is well detectable (in dependence on the experimental conditions such as the basicity of the used matrix) in both modes. This applies also to ESI MS.
10. It is very important to have a significant excess (about 100:1) of the detergent over the PL when using NMR techniques. Sodium cholate forms very small micelles which consist of just about four cholate molecules [23]. In the best case, one single PL molecule is entrapped in each of these small micelles—and this requires a considerable excess of the detergent.
11. It is not implied that under the used experimental conditions all lipids are completely extracted: some lipids may stick to the proteins that precipitate at the interphase between the aqueous and the organic layer and are, thus, lost. If protein-rich samples are investigated, higher ionic strengths, i.e., a high salt concentration (0.14 M or higher) are recommended in order to reduce the loss of lipids due to the binding to the protein. Precipitating large amounts of proteins by methanol and harsh centrifugation prior to lipid extraction might also be useful; however, a loss of desired lipids should be considered. Complete extraction of lipids from biological tissues is very difficult!

12. Unwanted lipid oxidation leads to the generation of peroxides which decay (under scission at the position of the original double) into aldehydes or carboxylic acids. This leads to characteristic mass differences which are summarized in [47]. For instance, oxidation of POPC (*m/z* 760.6 and 782.6) is reflected by the characteristic aldehyde peaks at *m/z* 650.5 and 672.5 [47].
13. The indicated extraction method is optimized for the analysis of PL. If more apolar lipids (e.g., TAG) are of interest, the moiety of chloroform should be enhanced. In the case of acidic PL, such as PS or PG and particularly phosphoinositides with several phosphate groups, slight acidification of the extraction mixture is recommended. In the latter case, the addition of acids is helpful in order to screen the charges by decreasing the pH.
14. Since there is no separation or purification prior to the NMR measurements great care should be taken to avoid the presence of phosphate. Phosphate is often used for the preparation of buffers, e.g., PBS—and well detectable by NMR (at about 2.2 ppm). Therefore, the use of a phosphate buffer to prepare the samples or to dilute a given cell suspension is strongly discouraged.
15. MALDI is (compared to ESI MS) a robust MS method and tolerates considerable salt contents within the sample, i.e., MALDI spectra may be recorded even at physiological salt concentration, although spectral quality is seriously reduced. Lipid extraction also helps to decrease the salt contents of the samples. Changes of the salt content may change the adduct patterns. Additionally, the intensities of typical interfering DHB peaks are also enhanced in the presence of salt.
16. One significant difference between MS and NMR is the extent of instrument contamination. While the NMR spectrometer is not in direct contact with the sample, MS instruments are always contaminated by the sample introduction. This is a particular problem in the case of ESI. In order to minimize the contamination, it is advisable to use strongly diluted samples which are typically less concentrated compared to MALDI MS.
17. The term "PC" should be used if there are exclusively diacyl species. "GPC" (glycerophosphorylcholine) is the more general term and also includes all types of ether lipids.

Acknowledgment

This study was supported by the German Research Council [DFG Schi 476/12-2, DFG Schi 476/16-1 and SFB 1052/Z3 (Project number 209933838)]. We would also like to thank all our colleagues who helped us performing the related experiments. We are particularly indebted to the MERCK company for the continuous support.

References

1. Schug ZT, Frezza C, Galbraith LC, Gottlieb E (2012) The music of lipids: how lipid composition orchestrates cellular behaviour. Acta Oncol 51:301–310
2. Mouritsen OG, Bagatolli LA (2015) Lipid domains in model membranes: a brief historical perspective. Essays Biochem 57:1–19
3. Schiller J, Arnold K (2002) Application of high resolution ^{31}P NMR spectroscopy to the characterization of the phospholipid composition of tissues and body fluids – a methodological review. Med Sci Monit 8:MT205–MT222
4. da Silva TF, Sousa VF, Malheiro AR, Brites P (2012) The importance of ether-phospholipids: a view from the perspective of mouse models. Biochim Biophys Acta 1822:1501–1508
5. Fahy E, Subramaniam S, Murphy RC, Nishijima M, Raetz CR, Shimizu T, Spener F, van Meer G, Wakelam MJ, Dennis EA (2009) Update of the LIPID MAPS comprehensive classification system for lipids. J Lipid Res 50 (Suppl):S9–S14
6. Pati S, Nie B, Arnold RD, Cummings BS (2016) Extraction, chromatographic and mass spectrometric methods for lipid analysis. Biomed Chromatogr 30:695–709
7. Bligh EG, Dyer WJ (1959) A rapid method of total lipid extraction and purification. Can J Biochem Physiol 3:911–917
8. Reis A, Rudnitskaya A, Blackburn GJ, Mohd Fauzi N, Pitt AR, Spickett CM (2013) A comparison of five lipid extraction solvent systems for lipidomic studies of human LDL. J Lipid Res 54:1812–1824
9. Abbott SK, Jenner AM, Mitchell TW, Brown SH, Halliday GM, Garner B (2013) An improved high-throughput lipid extraction method for the analysis of human brain lipids. Lipids 48:307–318
10. Sostare J, Di Guida R, Kirwan J, Chalal K, Palmer E, Dunn WB, Viant MR (2018) Comparison of modified Matyash method to conventional solvent systems for polar metabolite and lipid extractions. Anal Chim Acta 1037:301–315
11. Christie WW (2003) Lipid analysis. Oily Press, Bridgwater
12. Jurowski K, Kochan K, Walczak J, Barańska M, Piekoszewski W, Buszewski B (2017) Analytical techniques in lipidomics: state of the art. Crit Rev Anal Chem 47:418–437
13. Fuchs B, Süss R, Teuber K, Eibisch M, Schiller J (2011) Lipid analysis by thin-layer chromatography – a review of the current state. J Chromatogr A 1218:2754–2774
14. Li J, Vosegaard T, Guo Z (2017) Applications of nuclear magnetic resonance in lipid analyses: an emerging powerful tool for lipidomics studies. Prog Lipid Res 68:37–56
15. Fuchs B, Süß R, Schiller J (2010) An update of MALDI-TOF mass spectrometry in lipid research. Prog Lipid Res 49:450–475
16. Brügger B (2014) Lipidomics: analysis of the lipid composition of cells and subcellular organelles by electrospray ionization mass spectrometry. Annu Rev Biochem 83:79–98
17. Berger S, Braun S (2004) 200 and more NMR experiments: a practical course. Wiley, Weinheim
18. Popkova Y, Meusel A, Breitfeld J, Schleinitz D, Hirrlinger J, Dannenberger D, Kovacs P, Schiller J (2015) Nutrition-dependent changes of mouse adipose tissue compositions monitored by NMR, MS, and chromatographic methods. Anal Bioanal Chem 407:5113–5123
19. Guillén MD, Carton I, Goicoechea E, Uriarte PS (2008) Characterization of cod liver oil by spectroscopic techniques. New approaches for the determination of compositional parameters, acyl groups, and cholesterol from ^{1}H nuclear magnetic resonance and Fourier transform infrared spectral data. J Agric Food Chem 56:9072–9079
20. Pearce JM, Komoroski RA (2000) Analysis of phospholipid molecular species in brain by

$^{(31)}$P NMR spectroscopy. Magn Reson Med 44:215–223

21. Fuchs B, Müller K, Paasch U, Schiller J (2012) Lysophospholipids: potential markers of diseases and infertility? Mini Rev Med Chem 12:74–86
22. Schiller J, Müller M, Fuchs B, Arnold K, Huster D (2007) ^{31}P NMR spectroscopy of phospholipids: from micelles to membranes. Curr Anal Chem 3:283–301
23. Le Maire M, Champeil P, Moller JV (2000) Interaction of membrane proteins and lipids with solubilizing detergents. Biochim Biophys Acta 1508:86–111
24. Cole RB (2010) Electrospray and MALDI mass spectrometry: fundamentals, instrumentation, practicalities, and biological applications, 2nd edn. Wiley, Hoboken
25. Karas M, Glückmann M, Schäfer J (2000) Ionization in matrix-assisted laser desorption/ionization: singly charged molecular ions are the lucky survivors. J Mass Spectrom 35:1–12
26. Petković M, Schiller J, Müller M, Benard S, Reichl S, Arnold K, Arnhold J (2001) Detection of individual phospholipids in lipid mixtures by matrix-assisted laser desorption/ionization time-of-flight mass spectrometry: phosphatidylcholine prevents the detection of further species. Anal Biochem 289:202–216
27. Schröter J, Fülöp A, Hopf C, Schiller J (2018) The combination of 2,5-dihydroxybenzoic acid and 2,5-dihydroxyacetophenone matrices for unequivocal assignment of phosphatidylethanolamine species in complex mixtures. Anal Bioanal Chem 410:2437–2447
28. Sherma J, Fried B (2011) Thin-layer and high-performance thin-layer chromatographic analysis of biological samples. Adv Chromatogr 49:401–421
29. Gocan S (2002) Stationary phases for thin-layer chromatography. J Chromatogr Sci 40:538–549
30. Momchilova SM, Nikolova-Damyanova BM (2012) Advances in silver ion chromatography for the analysis of fatty acids and triacylglycerols −2001 to 2011. Anal Sci 28:837–844
31. Engel KM, Griesinger H, Schulz M, Schiller J (2019) Normal-phase versus reversed-phase thin-layer chromatography (TLC) to monitor oxidized phosphatidylcholines by TLC/mass spectrometry. Rapid Commun Mass Spectrom 33:60–65
32. Sun G, Yang K, Zhao Z, Guan S, Han X, Gross RW (2008) Matrix-assisted laser desorption/ionization time-of-flight mass spectrometric analysis of cellular glycerophospholipids enabled by multiplexed solvent dependent analyte-matrix interactions. Anal Chem 80:7576–7585
33. Fuchs B, Schiller J, Süss R, Schürenberg M, Suckau D (2007) A direct and simple method of coupling matrix-assisted laser desorption and ionization time-of-flight mass spectrometry (MALDI-TOF MS) to thin-layer chromatography (TLC) for the analysis of phospholipids from egg yolk. Anal Bioanal Chem 389:827–834
34. Luftmann H (2004) A simple device for the extraction of TLC spots: direct coupling with an electrospray mass spectrometer. Anal Bioanal Chem 378:964–968
35. Bresler K, Pyttel S, Paasch U, Schiller J (2011) Parameters affecting the accuracy of the MALDI-TOF MS determination of the phosphatidylcholine/lysophosphatidylcholine (PC/LPC) ratio as potential marker of spermatozoa quality. Chem Phys Lipids 164:696–702
36. Hillenkamp F, Peter-Katalinić J (2014) MALDI MS. Weinheim, Wiley-Blackwell
37. Schröter J, Popkova Y, Süß R, Schiller J (2017) Combined use of MALDI-TOF mass spectrometry and ^{31}P NMR spectroscopy for analysis of phospholipids. Methods Mol Biol 1609:107–122
38. Eibisch M, Fuchs B, Schiller J, Süß R, Teuber K (2011) Analysis of phospholipid mixtures from biological tissues by matrix-assisted laser desorption and ionization time-of-flight mass spectrometry (MALDI-TOF MS): a laboratory experiment. J Chem Educat 88:503–507
39. Fuchs B, Bischoff A, Süss R, Teuber K, Schürenberg M, Suckau D, Schiller J (2009) Phosphatidylcholines and -ethanolamines can be easily mistaken in phospholipid mixtures: a negative ion MALDI-TOF MS study with 9-aminoacridine as matrix and egg yolk as selected example. Anal Bioanal Chem 395:2479–2487
40. Zalata AA, Christophe AB, Depuydt CE, Schoonjans F, Comhaire FH (1998) The fatty acid composition of phospholipids of spermatozoa from infertile patients. Mol Hum Reprod 4:111–118
41. Teuber K, Schiller J, Jakop U, Lüpold S, Orledge JM, Blount JD, Royle NJ, Hoodless A, Müller K (2011) MALDI-TOF mass spectrometry as a simple tool to determine the phospholipid/glycolipid composition of sperm: pheasant spermatozoa as one selected example. Anim Reprod Sci 123:270–278
42. Lessig J, Gey C, Süss R, Schiller J, Glander HJ, Arnhold J (2004) Analysis of the lipid composition of human and boar spermatozoa by MALDI-TOF mass spectrometry, thin layer

chromatography and ^{31}P NMR spectroscopy. Comp Biochem Physiol B Biochem Mol Biol 137:265–277

43. White T, Bursten S, Federighi D, Lewis RA, Nudelman E (1998) High-resolution separation and quantification of neutral lipid and phospholipid species in mammalian cells and sera by multi-one-dimensional thin-layer chromatography. Anal Biochem 258:109–117
44. Gouilleux B, Christensen NV, Malmos KG, Vosegaard T (2019) Analytical evaluation of low-field ^{31}P NMR spectroscopy for lipid analysis. Anal Chem 91:3035–3042
45. Wu J, Teuber K, Eibisch M, Fuchs B, Schiller J (2011) Chlorinated and brominated phosphatidylcholines are generated under the influence of the Fenton reagent at low pH – a MALDI-TOF MS study. Chem Phys Lipids 164:1–8
46. Griesinger H, Fuchs B, Süß R, Matheis K, Schulz M, Schiller J (2014) Stationary phase thickness determines the quality of thin-layer chromatography/matrix-assisted laser desorption and ionization mass spectra of lipids. Anal Biochem 451:45–47
47. Fuchs B, Schiller J, Süß R, Nimptsch A, Schürenberg M, Suckau D (2009) Capabilities and disadvantages of combined matrix-assisted laser desorption/ionization time-of-flight mass spectrometry (MALDI-TOF MS) and high-performance thin-layer chromatography (HPTLC): analysis of egg yolk lipids. J Planar Chromatogr 22:35–42

Chapter 6

Analysis of Sterols by Gas Chromatography–Mass Spectrometry

Ashutosh Singh, Sana Akhtar Usmani, Khushboo Arya, and Nitin Bhardwaj

Abstract

Sterols are a major component of cell membranes among all biological systems including bacteria, plant, fungi, and mammals. While the essential carbon skeleton found in all sterol-like structures is the sterane (cyclopentanoperhydrophenanthrenes) ring, there are minor variations which make each structure unique. These include hydroxylations, methylations, ketone groups, double bonds, etc. These structures play specific roles in the biological membranes. Earlier, sterols could only be detected using traditional methods like thin-layer chromatography or UV–vis spectrophotometry. However, these techniques are unable to accurately differentiate between these closely related sterol structures. Therefore, it becomes essential to develop new and sensitive methods for accurate quantification of sterols. In the last few decades, research on gas chromatography–mass spectrometry (GCMS)-based sterol structure determination and quantification has been on the rise. In this chapter, we have discussed some basic background of GCMS and its application in the absolute quantification of sterols using some examples.

Keywords Gas chromatography, Mass spectrometry, Sterols

Abbreviations

BHT	Butylated hydroxytoluene
BSTFA-TMCS	(*N,O*-Bis(trimethylsilyl)trifluoroacetamide-trimethylchlorosilane)
CI	Chemical ionization
EI	Electron impact
EIC	Extracted ion chromatogram
GC	Gas chromatography
i.s.	Internal standard
m/z	Mass-to-charge ratio
MS	Mass spectrometry
RT	Retention time
TIC	Total ion chromatogram
TLC	Thin-layer chromatography
TMS	Trimethylsilyl

Rajendra Prasad and Ashutosh Singh (eds.), *Analysis of Membrane Lipids*, Springer Protocols Handbooks, https://doi.org/10.1007/978-1-0716-0631-5_6,

1 Introduction

Gas chromatography–mass spectrometry (GCMS) is a highly specific, sensitive, and accurate technique to separate and analyze small volatile compounds. Separation of analytes on GCMS is based mainly on two critical features namely the volatility and polarity. Essentially the setup relies on the separation of thermally stable volatile components on the GC and subsequent detection of the separated component on the MS based on the mass-to-charge ratio (*m/z*) [1].

Briefly, sample solution is injected into the GC inlet where it vaporizes, and then the vaporized analytes are guided into the GC column with the help of an inert mobile carrier gas (He, helium). The difference between the chemical properties of the solute and their relative affinity toward the stationary phase of the column allows them to separate as the sample moves through the length of the column. Different molecules elute at different times from the column and then enter into the mass spectrometer detector. In MS, the samples are ionized using electron (EI) or chemical (CI) ionization sources. In electron ionization, a beam of electron ionizes the sample resulting in the loss of one electron resulting in formation of a radical carbocation. This method is most commonly used. In chemical ionization, a reagent gas methane is introduced which interacts with the analytes and causes their ionization. The ionized molecules are accelerated through the instruments mass analyzer which is a quadruple or an iron trap. Now ions are separated according to their *m/z*. Different peaks appear according to their *m/z* ratio and a spectrum is generated [1, 2].

Of recently, GCMS platforms have been extensively used in the following:

1. Environmental research: Detection of chloroform, polycyclic aromatic hydrocarbons, gasoline, dioxins, dibenzofurans, pesticides, herbicides, phenols, sulfur [3].
2. Forensic research: Detection of doping compounds like steroids, narcotics, alcohol, and drug residues [3, 4].
3. Agricultural research: Pesticide detection [5].
4. Energy research: Detection of aromatic solvents, sulfur, impurities in polypropylene, ethylene, gas oil, diesel oil, polythene grafted polymers, etc. [6, 7].
5. Basic academic research: Steroids, fatty acyls, hormones, etc. are among the thousands of compounds that can be analyzed by GCMS [8, 9].

In 1960s, the use of MS combined with GC for the analysis of sterols and steroid was recognized. Since then this technique has seen major improvements in the field of analyses of complex sterol mixtures. The technique has several advantages like high sensitivity of detection and high resolution of closely related complex structures on fused-silica bonded stationary phase capillary GC columns that can be operated on high temperatures (for example (5%-phenyl)-methylpolysiloxane-coated columns) which makes it easier to analyze sterols or steryl conjugates, powerful computers that make analyses part easy, and availability of large libraries of comparative compound spectra which allow quick identification even without the standards [10].

Sterols are found in all eukaryotic organisms as an important membrane component, where they regulate the fluidity and permeability of the membrane [11]. However, prokaryotic systems do not contain sterols. In yeasts, the principal sterol is ergosterol which maintains the integrity of the membrane. Sterol occurs in plants and animals and is an essential component. Major plant sterols which are available in edible material are sitosterol, stigmasterol, and campesterol [12]. They are found in cereals, nuts, vegetables, legumes, unrefined vegetable oils, etc. Plant sterols also play a key role in cellular signaling mechanisms. Cholesterol is an important constituent of animal food products and is often related with coronary heart disease. Accumulation of oxysterols can cause different diseases such as inflammatory bowel disease, atherosclerosis, and neurodegenerative disorders [13]. Structures of some common sterols are shown in Fig. 1 as their trimethylsilyl (TMS) derivatives. Although different sterol structures are found in different organisms, but their backbone skeleton component remains the same viz. cyclopentanoperhydrophenanthrene. The minor variations in different sterol structures cannot be detected by routine techniques like thin-layer chromatography (TLC) or spectrophotometry.

The use of GC and GCMS allows us to separate multiple sterol structures in a single sample run. Though there are many advantages of GCMS, this technique has some limitations also. For example, sample analysis is often time consuming, mostly due to lack of appropriate standards. Also, EI is a hard ionization technique which often results in total loss of the molecular ion, making analysis rather difficult. Nonetheless, these techniques are widely used for sterol analysis due to their chromatographic resolving capacity, their robustness, and the relatively low cost of acquiring and operating the instruments [10, 14]. Below we have discussed some key aspects of sterol analysis by GCMS.

2 Sample Preparation

Sterols can be directly extracted from the samples; however in the era where the global lipidomes are explored, total lipid extractions are done prior to the enrichment of any specific lipid group like sterols. Several protocols have been adopted for the total lipid

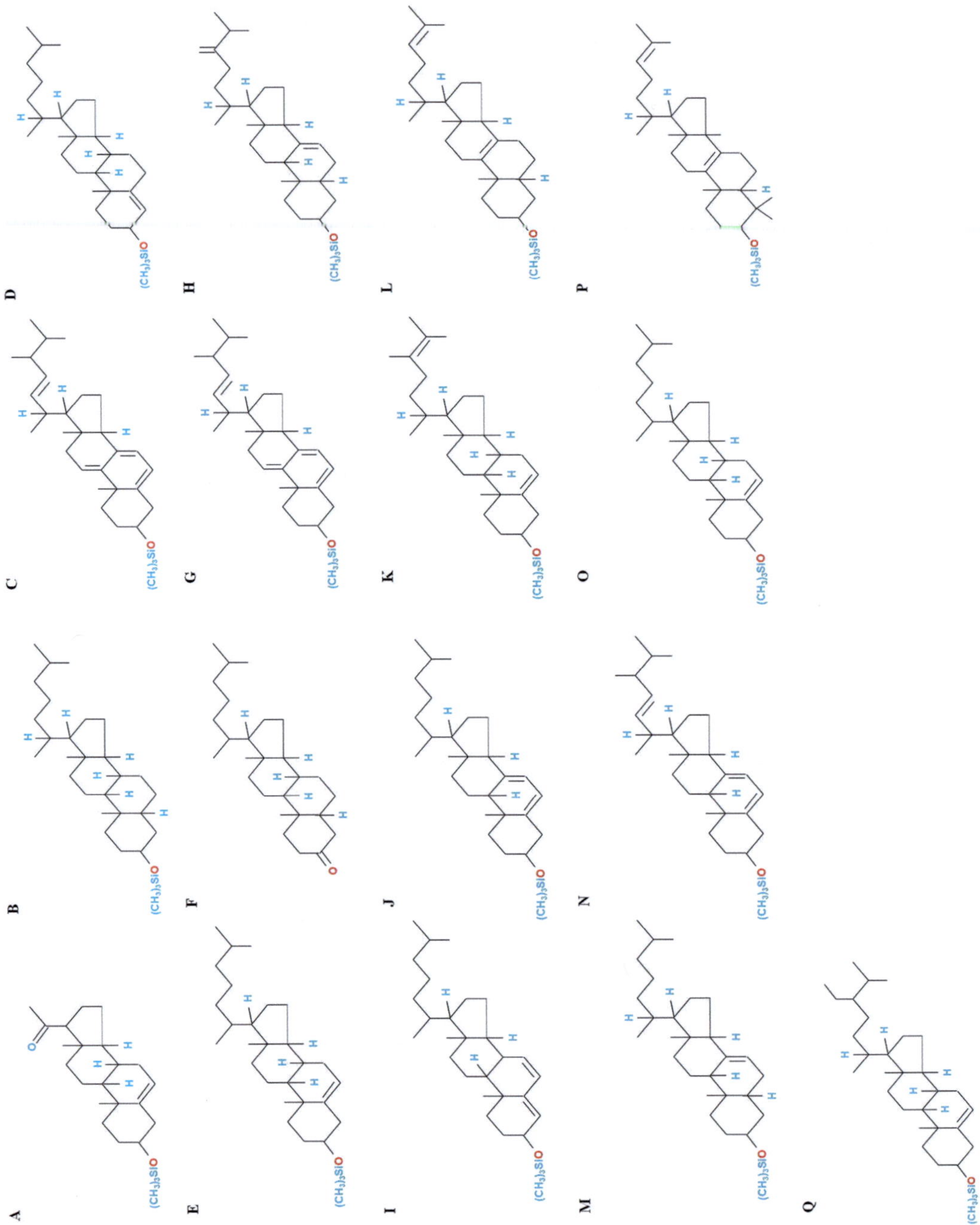
A
B
C
D
E
F
G
H
I
J
K
L
M
N
O
P
Q

Fig. 1 TMS-derivatized structures of various sterols. The figure presents structures of TMS derivatives of: (**a**) pregnenolone [3β-Hydroxypregn-5-en-20-one]; (**b**) coprostanol [5β-Cholestan-3β-ol]; (**c**) dehydroergosterol [Ergosta-5,7,9(11),22-tetraen-3β-ol]; (**d**) allocholesterol [Cholest-4-en-3β-ol]; (**e**) cholesterol [Cholest-5-en-3-ol]; (**f**) 5β-cholestanone [5β-Cholestan-3-one]; (**g**) dihydrocholesterol [Cholestan-3β-ol]; (**h**) episterol [24-methylene-cholest-7-en-3β-ol]; (**i**) cholesta-4,6-dien-3-ol [cholesta-4,6-dien-3-ol, (3.beta.)-]; (**j**) 7-dehydrocholesterol [cholesta-5,7-dien-3β-ol]; (**k**) desmosterol [cholest-5,24-dien-3β-ol]; (**l**) zymosterol [5α--cholesta-8,24-dien-3β-ol]; (**m**) lathosterol [cholest-7-en-3β-ol]; (**n**) erogosterol [ergosta-5,7,22E-trien-3β-ol]; (**o**) cholestenone [cholest-4-en-3-one]; (**p**) lanosterol [lanosta-8,24-dien-3β-ol]; (**q**) sitosterol [Stigmast-5-en-3β-ol]. The structures have been adapted from www.lipidmaps.org

extraction. In yeasts, sample preparation requires accurate cell count, best done by hemocytometer, and then the cells are suspended in desired buffer system and homogenized using zirconia beads (0.5 mm) by vortexing/sonication or by Daum's homogenizer. Usually, the ratio of beads:suspended sample is kept at 1:2 (v/v). Ratios may vary depending upon the type of sample. An internal standard (i.s.) is added to each sample prior to lipid extraction to keep in check the extraction efficiencies and provide accurate quantification (*see* **Note 1**). For example, ergosterol can be added as i.s. for plant and mammalian sterol estimations. Cholesterol is added as i.s. for yeast sterol analyses. Further MS grade solvents containing 0.01% butylated hydroxytoluene (BHT) is preferred for lipid extraction (*see* **Note 3**).

Several approaches can be employed for the total lipid extraction:

Folch Extraction [15]

1. Homogenized 1 g tissue/sample (with an assumption that it has the specific gravity of water) is mixed with 20 mL of chloroform:methanol (2:1; v/v).
2. The mixture is vortexed well and kept at 37 °C for at least 2 h to overnight.
3. Sample is then washed with 0.9% NaCl to remove any polar impurities and phase separation (*see* **Note 2**). The ratio of chloroform:methanol:water is kept at 8:4:3 (v/v/v).
4. Sample is then centrifuged at 820 × *g* for 20 min.
5. The lower hydrophobic layer is aspirated using the Pasteur pipette into a separate tube.
6. Lipid extract is then dried in SpeedVac, flushed with N_2 (*see* **Note 6**), and kept at −20 °C until further analysis.

Bligh and Dyer's Method [16]

1. Homogenized 1 g biological sample (with an assumption that it contains at least 80% water and 1% lipid) is mixed with 3 mL of chloroform:methanol (1:2; v/v) resulting into the chloroform:methanol:water (1:2:0.8; v/v/v).
2. The mixture is vortexed well and kept at 37 °C for at least 2 h.
3. Add 1 mL chloroform and 1 mL water, vortex well (twice for 30 s). This results in final proportions of chloroform:methanol: water of 2:2:1.8 (v/v/v).

4. The sample mixture is kept at 37 °C for at least 2 h to overnight.
5. The lower hydrophobic layer is aspirated using the Pasteur pipette into a separate tube.
6. Sample is then washed with 0.9% NaCl to remove any polar impurities and phase separation.
7. Sample is then centrifuged at 820 × *g* for 20 min.
8. Again, the lower hydrophobic layer is aspirated using the Pasteur pipette into a separate tube.
9. Lipid extract is then dried in SpeedVac, flushed with N_2, and kept at −20 °C until further analysis.

Two-step lipid extraction [17]

1. Homogenized samples (~200 μL containing 1 OD cells) are extracted with 1 mL chloroform:methanol (17:1; v/v).
2. The mixture is vortexed well and kept at 37 °C for at least 2 h.
3. The lower hydrophobic layer (17:1; v/v) is aspirated using the Pasteur pipette into a separate tube. Lipid extract is then dried under N_2 and kept at −20 °C until further analysis. This extract is sterol rich.
4. The upper aqueous layer is extracted with 1 mL chloroform: methanol (2:1; v/v).
5. The lower hydrophobic layer (2:1; v/v) is aspirated using the Pasteur pipette into a separate tube. Lipid extract is then dried in SpeedVac, flushed with N_2, and kept at −20 °C until further analysis.

Mandala extraction [18]

1. 5×10^8 cells are homogenized with glass beads in 1.5 mL ethanol: dH_2O:diethylether:pyridine:NH_4OH (15:15:5:1:0.018; v/v) by alternate vortexing and sonication of 30 s.
2. Sample is kept at 60 °C water bath for 15 min.
3. Sample is vortexed and sonicated for 30 s each and kept at 60 °C water bath for another 15 min.
4. Sample is then sonicated and centrifuged at 1300 × *g* for 10 min at 4 °C.
5. Dry the supernatant in SpeedVac. Avoid over-drying of samples.

- Samples at this stage can be flushed with N_2 and stored at −20 °C.
- Samples from hereon are processed with minor modification of Bligh and Dyer's method.

6. To the dry pellet, 2 mL methanol is added followed by severe vortexing and sonication, until the pellet is completely dissolved or in phase.
7. Sample is kept at 37 °C for 1 h with vortexing (twice for 30 s).
8. Sample is then centrifuged at 1300 × *g* for 10 min at 25 °C.
9. The supernatant is transferred to a fresh tube using the Pasteur pipette.
10. Add 1 mL chloroform and 1 mL water, vortex well (twice for 30 s).
11. Sample is then centrifuged at 1300 × *g* for 5 min at 25 °C.
12. The lower hydrophobic layer is aspirated using the Pasteur pipette into a separate tube.
13. Lipid extract is then dried in SpeedVac, flushed with N_2, and kept at −20 °C until further analysis.

Base Hydrolysis of Lipid Extracts: Mild alkaline hydrolysis is required to remove any polar lipids from the lipid extracts, specifically phosphoglycerides [19]. This allows enrichment of lipid groups like sterols and sphingolipids in the sample. Therefore, this step is mandatory after total lipid extraction.

1. To the dry lipid pellet, add 1 mL chloroform and 1 mL 0.6 M KOH in methanol.
2. Vortex well and leave at 25 °C for 1 h.
3. Then add 650 μL 1 M HCl and 250 μL water to the sample.
4. Vortex the sample thoroughly and centrifuge at 1300 × *g* for 10 min at 25 °C.
5. The lower hydrophobic layer is aspirated using the Pasteur pipette into a separate tube.
6. The base-hydrolyzed lipid extract is then dried in SpeedVac, flushed with N_2, and kept at −20 °C until further analysis.

Direct Extraction of Sterols: One of the older methods of sterol analysis involves direct extraction of sterols from biological samples using alkaline hydrolysis [20].

1. About 200 mg (dry weight) of biological sample is mixed with 3 mL methanol, 2 mL 0.5% pyrogallol, and 2 mL 60% KOH, vortex well.
2. Sample is kept at 60 °C water bath for 2 h.
3. Cool the sample. Extract the sample with 5 mL *n*-heptane by vortexing. Centrifuge at 1300 × *g* for 10 min at 25 °C to give phase separation. Remove the upper layer of *n*-heptane into a separate tube.
4. Re-extract the aqueous phase two more times with 5 mL *n*-heptane and combine all three *n*-heptane layers.
5. Dry the sample in SpeedVac (or evaporate in water bath at 55 °C), flush with N_2, and keep at −20 °C until further analysis.

Derivatization of Sterols: Sterols are non-volatile entities and require chemical derivatization to make them volatile [21]. In this regard, the TMS derivatization is highly recommended and allows efficient volatization. It is important to note that the derivatization step be only performed if the analysis is to be performed within 48 h.

1. Add 100 μL of *N*,*O*-bis(trimethylsilyl)trifluoroacetamide + trimethylchlorosilane (BSTFA + TMCS, 99:1, v/v) to the dry base-hydrolyzed lipid extract.
2. Heat at 90 °C for 2 h for derivatization to complete.
3. Cool down the sample, centrifuge, and transfer the clear supernatant into GC sample vials.
4. Add 50 μL of *n*-hexane to sample mix properly by syringe. The sample is now ready to be analyzed.

In another derivatization method, the base-hydrolyzed dry lipid extract is dissolved in 50 μL pyridine, and to this 50 μL BSTFA is added and mixed well by pipetting. Samples can now be analyzed on GC.

3 GCMS Analysis

Column: For sterol analysis, the most preferred and readily available GC column is DB-5 ms (Agilent Technologies) which features the non-polar phenyl arylene polymer (very similar to (5%-phenyl)-methylpolysiloxane) matrix. Although these columns come in different dimensions, the one used most commonly is DB-5 ms GC Column, 30 m (length), 0.25 mm (inner diameter), 0.25 μm (film). Other common examples of the general purpose columns with (5%-phenyl)-methylpolysiloxane include Rtx-5MS (Restek), ZB-5MS (Phenomenex), etc.

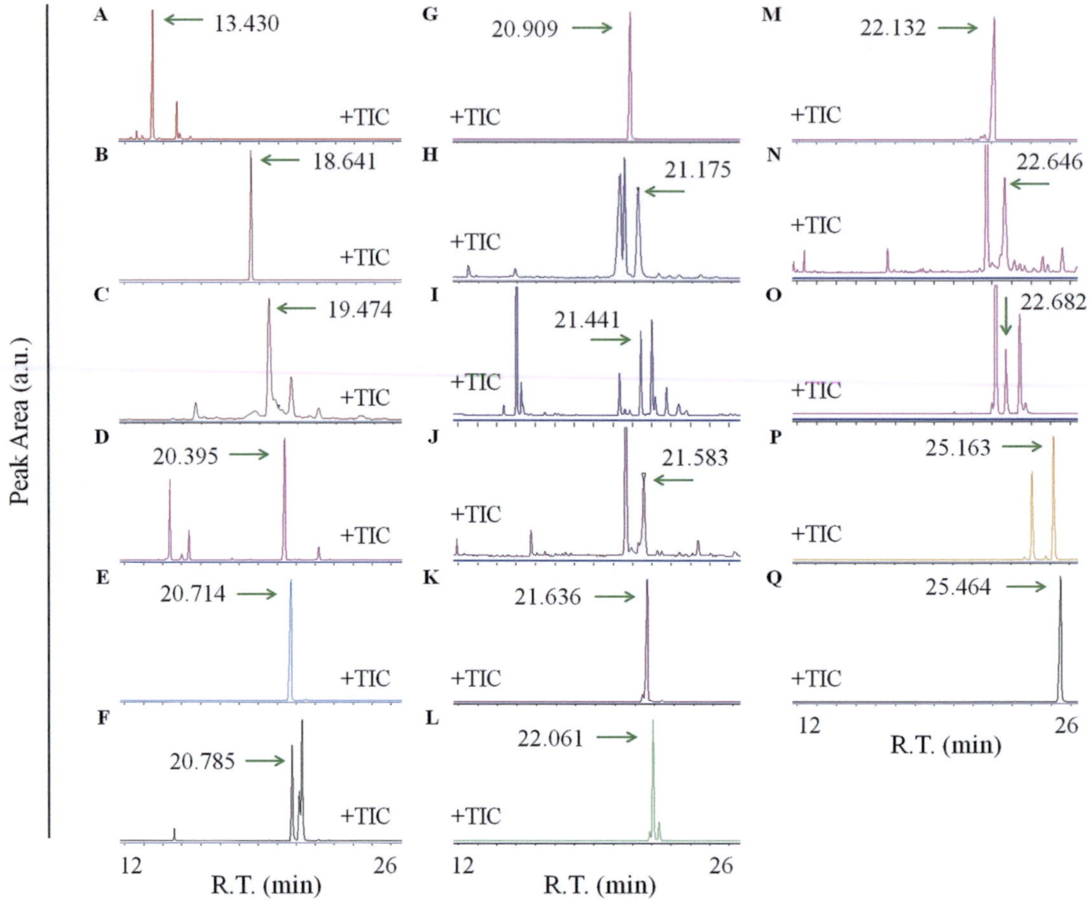

Fig. 2 TICs of TMS-derivatized sterol structures obtained by GCMS. The figure presents TICs of TMS derivatives of: (**a**) pregnenolone; (**b**) coprostanol; (**c**) dehydroergosterol; (**d**) allocholesterol; (**e**) cholesterol; (**f**) 5β-cholestanone; (**g**) dihydrocholesterol; (**h**) episterol; (**i**) cholesta-4,6-dien-3-ol; (**j**) 7-dehydrocholesterol; (**k**) desmosterol; (**l**) zymosterol; (**m**) lathosterol; (**n**) erogosterol; (**o**) cholestenone; (**p**) lanosterol; (**q**) sitosterol. The obtained retention times for each sterol species are marked with green arrows

GCMS Parameters: Sterols are analyzed using the initial column temperature of 100 °C with hold of 5 min, ramped at 35 °C/min to 240 °C, and further ramped at 3 °C/min to 305 °C with hold of 5 min. All EI-mass spectra were recorded at 70 eV with MS ion source temperature of 230 °C and MS quadruple temperature of 150 °C. Injection volume of 1 μL in splitless mode is kept. Split mode may be applied for more concentrated samples. The inlet port temperature and MSD transfer line temperature are kept at 290 and 280 °C. The flow rate of carrier gas (helium) of 1 mL/min is optimal. The solvent cutoff time for hexane was kept at 11 min [19, 22]. Of course one can change these parameters to increase resolution or decrease run time and so on.

Figure 2 represents the total ion chromatogram (TIC) and the retention times (RT) of various TMS-derivatized sterol structures

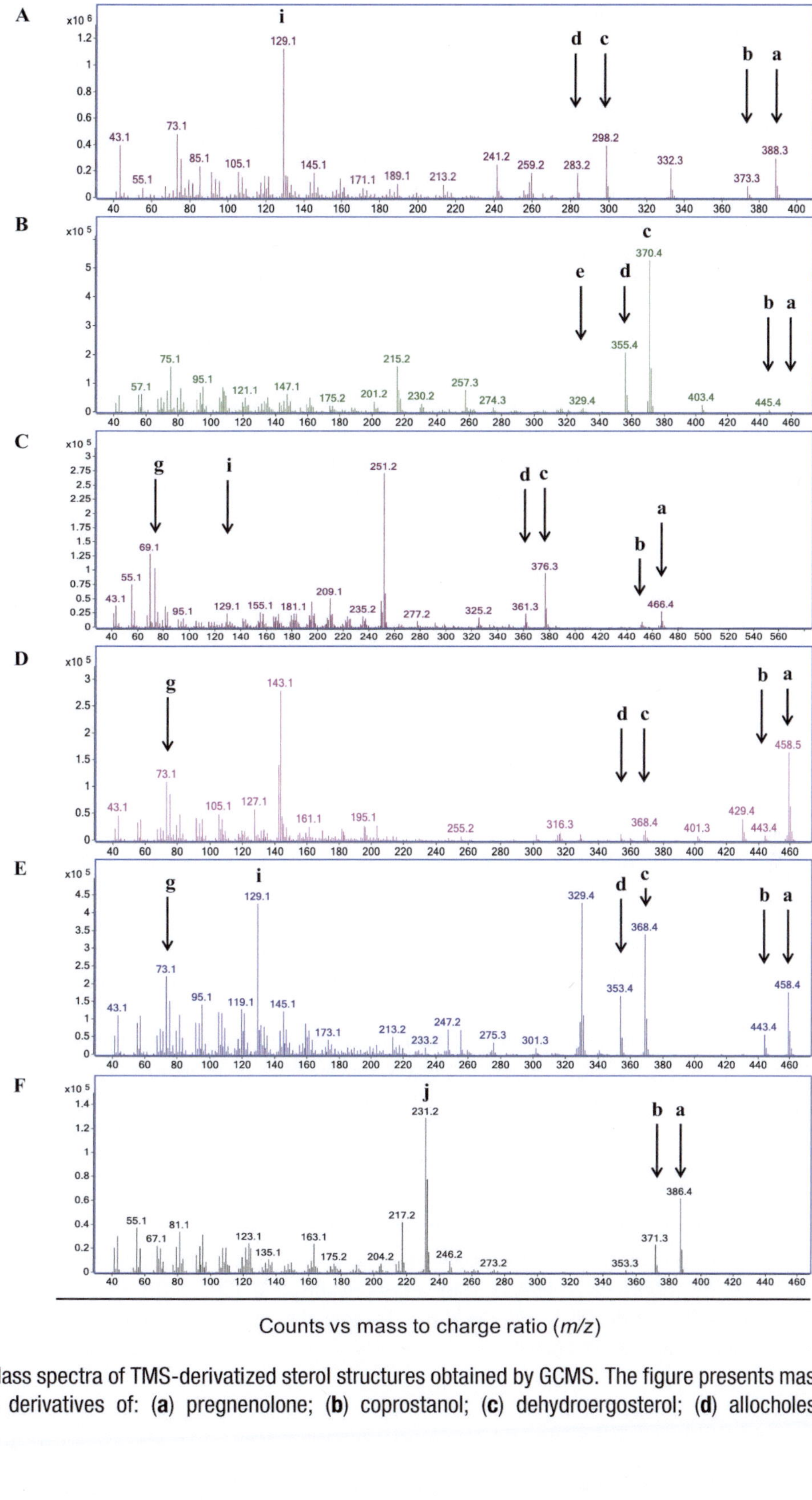

Fig. 3 Mass spectra of TMS-derivatized sterol structures obtained by GCMS. The figure presents mass spectra of TMS derivatives of: (**a**) pregnenolone; (**b**) coprostanol; (**c**) dehydroergosterol; (**d**) allocholesterol; (**e**)

depicted in Fig. 1. GC allows a clear separation of closely related sterol structures. We can clearly see that all sterol structures can be resolved in a narrow window of 13–26 min. For clarity each sterol structure has been depicted as a separate run. It is important to note that some sterol standards showed more than one peak in the chromatogram which implicates that either the sterol standard is impure or the sample preparation affects the sample integrity. Therefore a thorough MS analysis is mandated for such structures. Figures 3, 4, and 5 show the fragmentation pattern of various TMS-derivatized sterol structures. In the mass spectrum, apart from the molecular ion peak (M^+), one can observe characteristic peaks like M–CH_3, M–$(CH_3)_3SiOH$, M–$(CH_3)_3SiOH$–CH_3, M–$(CH_3)_3SiO^+$=CH–CH_2–CH_3, M–$(CH_3)_3SiOH$-side chain, $(CH_3)_3Si$, $(CH_3)_3SiO^+$=CH–CH_2–CH_3, etc. [23]. Once we have a clean spectrum, specific *m/z* ions can be used to identify and quantify specific sterol structure.

Figure 6 shows the extracted ion chromatograms (EIC) using a single ion. For example, pregnenolone can be detected using the *m/z* of 373.3 [22], or ergosterol can be detected using the *m/z* of 363.3 [19], etc. We can clearly see that:

1. The retention times match tightly with those observed in TICs.
2. The EICs are much neater with far less background peaks compared to TICs.
3. This single ion can be used to distinguish between closely related sterol structures.

One can use the area under the curve of these EICs to generate calibration curves. Figure 7 shows calibration curves for EICs of different sterol structures analyzed by GCMS. These plots have been normalized for internal standards. These curves remain linear between 7 ng and 1.4 μg on column. The lower limit of detection is about 70 ng on column for most sterol structures. However a lower detection limit of as low as 7 ng is observed for cholesterol, copratsterol, episterol, and lathosterol. These calibration curves are used to determine the concentration of any specific sterol structure in an unknown sample. The normalized data can then be represented as the amount of sterol/ OD cells, or /mg dry lipid weight, etc. (*see* **Note 7**).

Therefore the GCMS-based analysis is an accurate and sensitive method for the detection of various sterol structures [19, 21–25].

Fig. 3 (continued) cholesterol; (**f**) 5β-cholestanone. Here, a, b, c, d, e, f, g, h, i, j represent M^+, M–CH_3, M–$(CH_3)_3SiOH$, M–$(CH_3)_3SiOH$–CH_3, M–$(CH_3)_3SiO^+$=CH–CH_2–CH_3, M–$(CH_3)_3SiOH$-side chain, $(CH_3)_3Si$, $(CH_3)_3SiO^+$=CH–CH_2–CH_3, $(CH_3)_3SiO^+$=CH–CH=CH_2, Tri-aromatic steroid structure, respectively. The y-axis represents the signal intensity (a.u.)

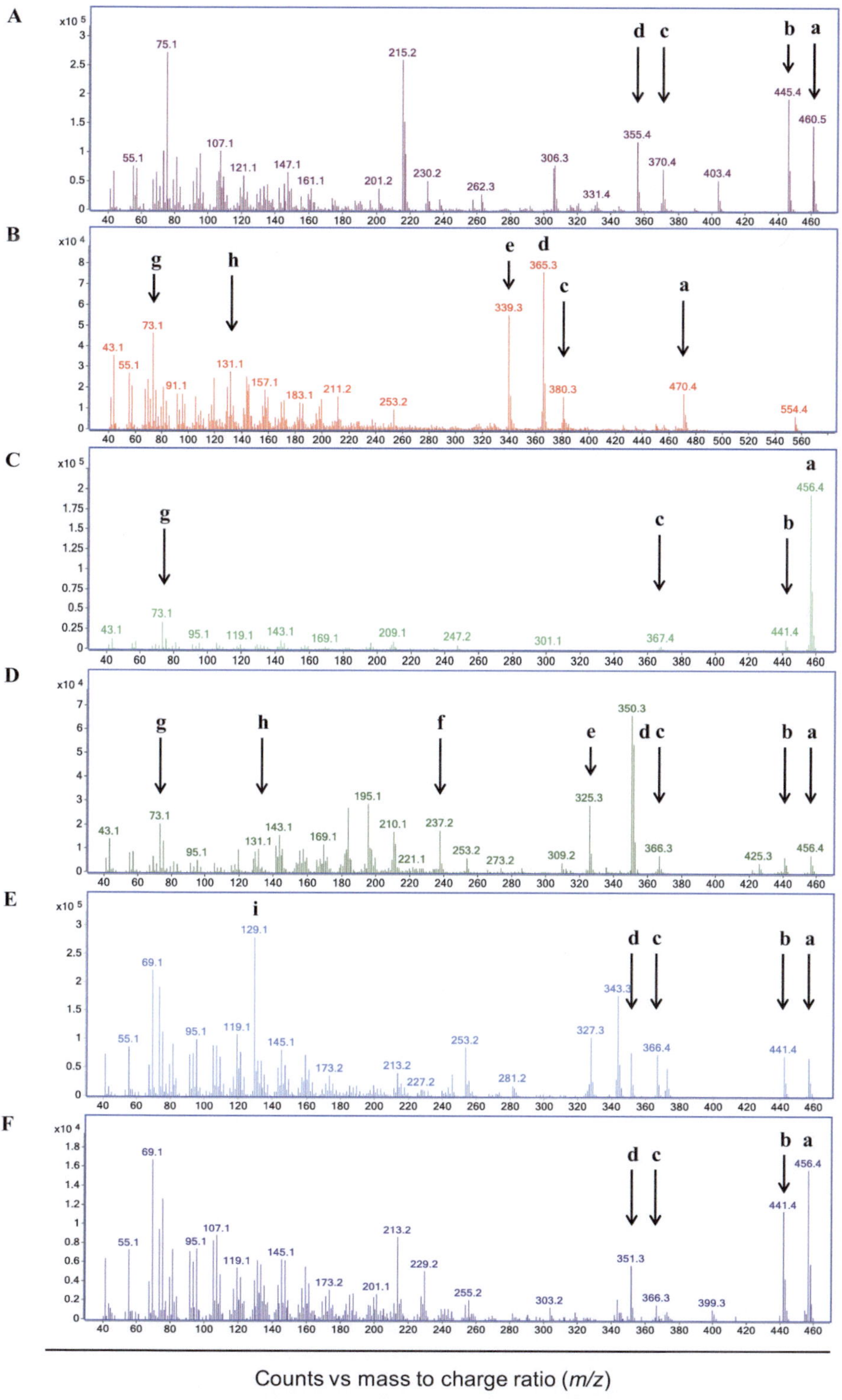

Fig. 4 Mass spectra of TMS-derivatized sterol structures obtained by GCMS. The figure presents mass spectra of TMS derivatives of: (**a**) dihydrocholesterol; (**b**) episterol; (**c**) cholesta-4,6-dien-3-ol; (**d**) 7-dehydrocholesterol; (**e**) desmosterol; (**f**) zymosterol. Here, a, b, c, d, e, f, g, h, i, j represent M^+, $M–CH_3$, $M–(CH_3)_3SiOH$, $M–(CH_3)_3SiOH–CH_3$, $M–(CH_3)_3SiO^+{=}CH–CH_2–CH_3$, $M–(CH_3)_3SiOH$-side chain, $(CH_3)_3Si$, $(CH_3)_3SiO^+{=}CH–CH_2–CH_3$, $(CH_3)_3SiO^+{=}CH–CH{=}CH_2$, Tri-aromatic steroid structure, respectively. The y-axis represents the signal intensity (a.u.)

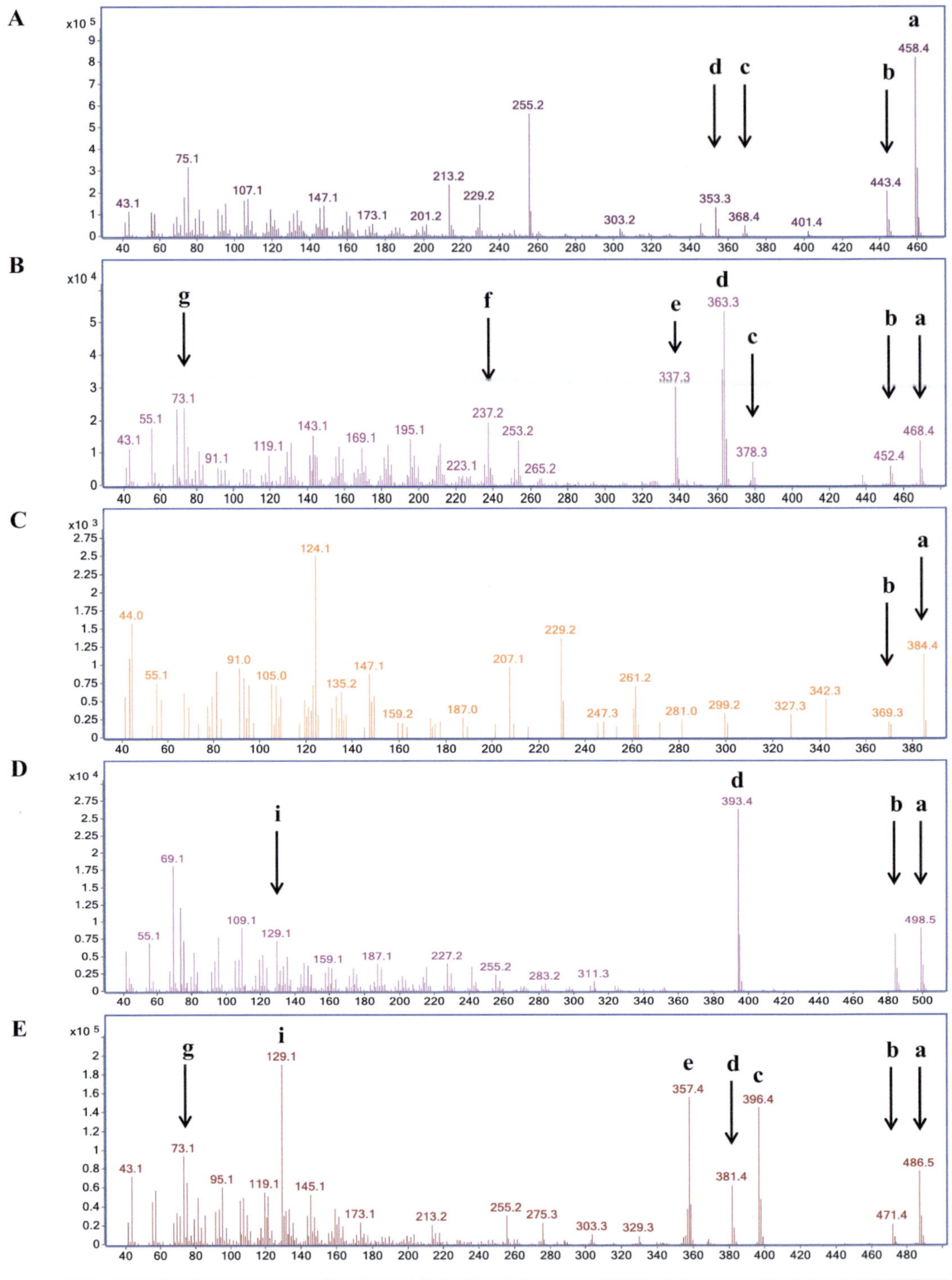

Counts vs mass to charge ratio (*m/z*)

Fig. 5 Mass spectra of TMS-derivatized sterol structures obtained by GCMS. The figure presents mass spectra of TMS derivatives of: (**a**) lathosterol; (**b**) erogosterol; (**c**) cholestenone; (**d**) lanosterol; (**e**) sitosterol. Here, a, b, c, d, e, f, g, h, i, j represent M^+, M–CH_3, M–$(CH_3)_3SiOH$, M–$(CH_3)_3SiOH$–CH_3, M–$(CH_3)_3SiO^+$=CH–CH_2–CH_3, M–$(CH_3)_3SiOH$-side chain, $(CH_3)_3Si$, $(CH_3)_3SiO^+$=CH–CH_2–CH_3, $(CH_3)_3SiO^+$=CH–CH=CH_2, Tri-aromatic steroid structure, respectively. The y-axis represents the signal intensity (a.u.)

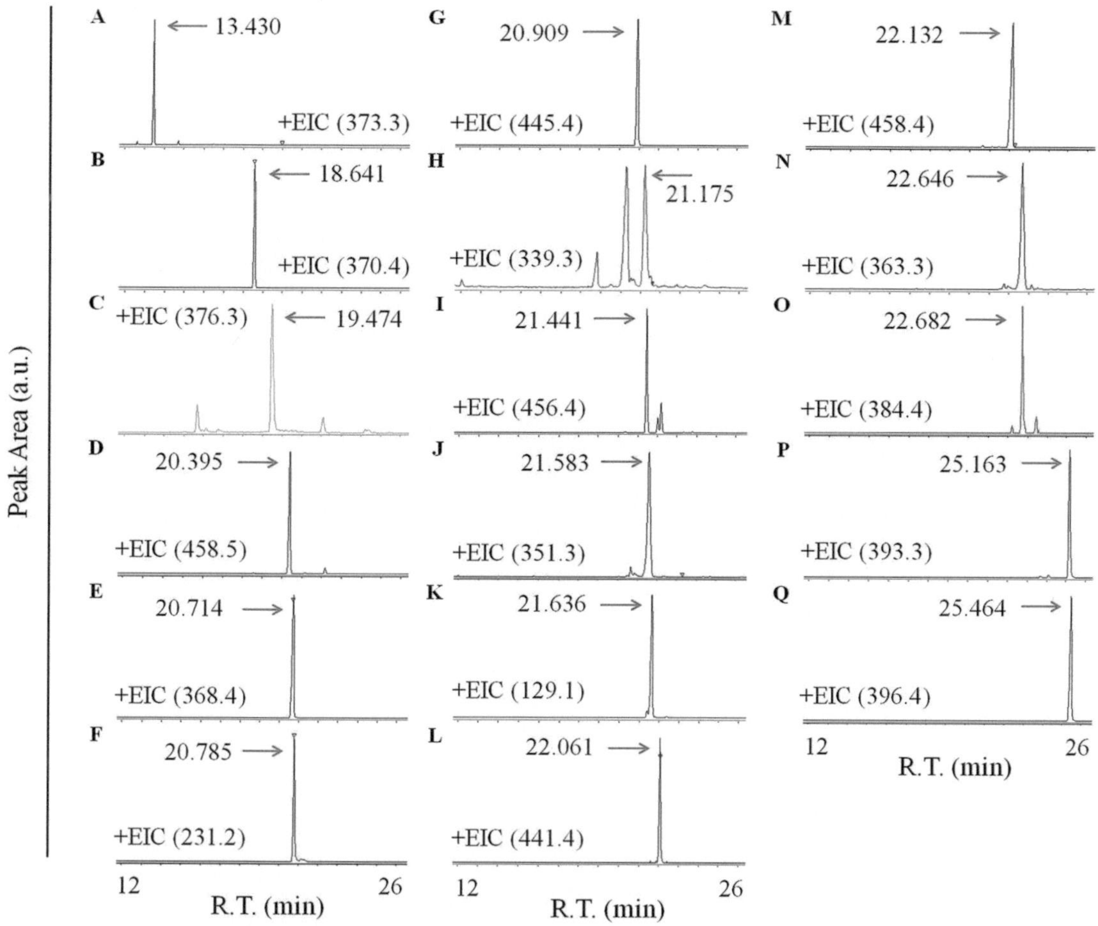

Fig. 6 EICs of TMS-derivatized sterol structures obtained by GCMS. The figure presents EICs of TMS derivatives of: (**a**) pregnenolone [373.3]; (**b**) coprostanol [370.4]; (**c**) dehydroergosterol [376.3]; (**d**) allocholesterol [458.5]; (**e**) cholesterol [368.4]; (**f**) 5β-cholestanone [231.2]; (**g**) dihydrocholesterol [445.4]; (**h**) episterol [339.3]; (**i**) cholesta-4,6-dien-3-ol [456.4]; (**j**) 7-dehydrocholesterol [351.3]; (**k**) desmosterol [129.1]; (**l**) zymosterol [441.4]; (**m**) lathosterol [458.4]; (**n**) erogosterol [363.3]; (**o**) cholestenone [384.4]; (**p**) lanosterol [393.3]; (**q**) sitosterol [396.4]. The obtained retention times for each sterol species are marked with green arrows

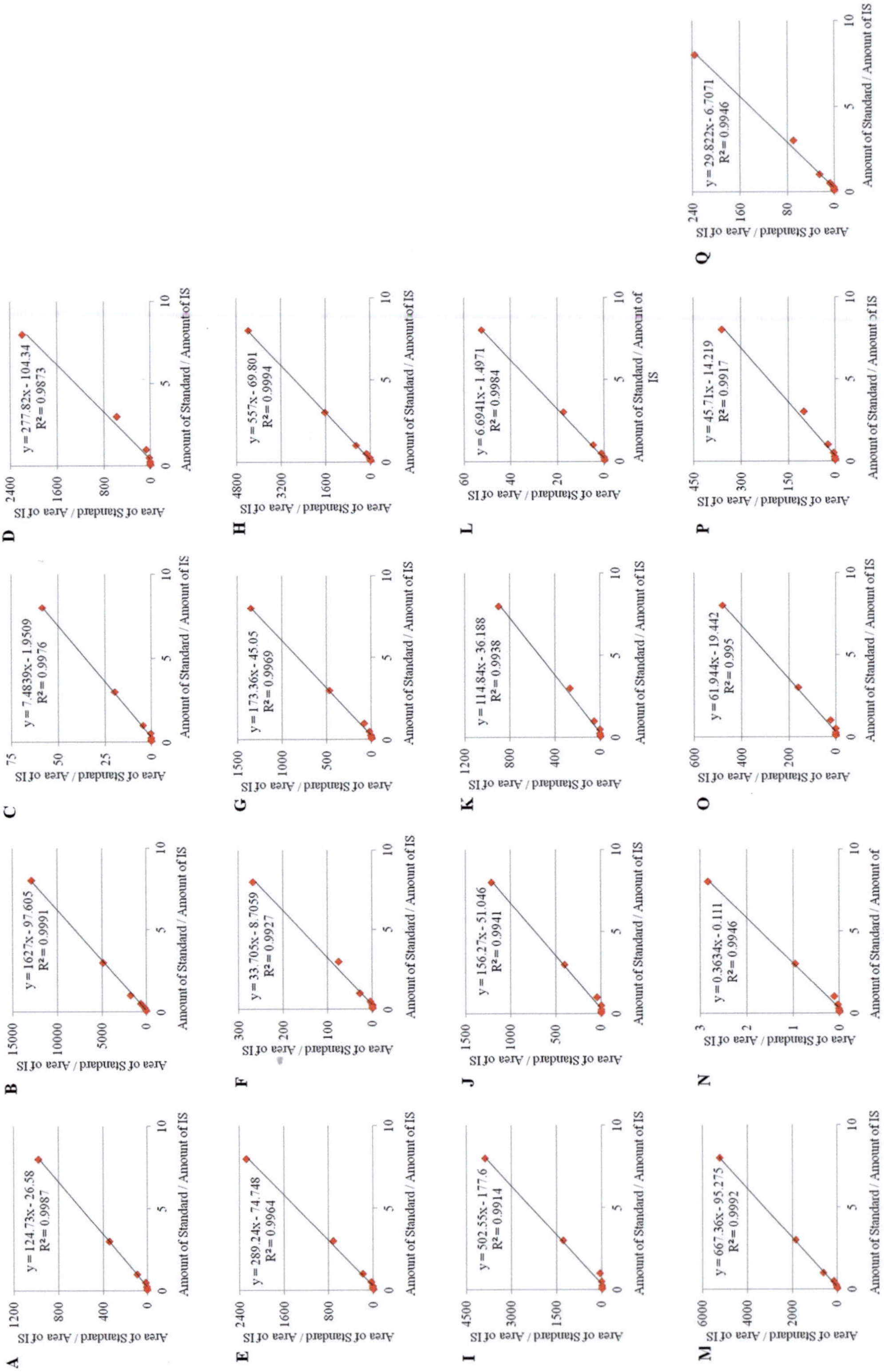
A
y = 124.73x - 26.58
R² = 0.9987
B
y = 1627x - 97.605
R² = 0.9991
C
y = 7.4839x - 1.9509
R² = 0.9976
D
y = 277.82x - 104.34
R² = 0.9873
E
y = 289.24x - 74.748
R² = 0.9964
F
y = 33.705x - 8.7059
R² = 0.9927
G
y = 173.36x - 45.05
R² = 0.9969
H
y = 557x - 69.801
R² = 0.9994
I
y = 502.55x - 177.6
R² = 0.9914
J
y = 156.27x - 51.046
R² = 0.9941
K
y = 114.84x - 36.188
R² = 0.9938
L
y = 6.6941x - 1.4971
R² = 0.9984
M
y = 667.36x - 95.275
R² = 0.9992
N
y = 0.3634x - 0.111
R² = 0.9946
O
y = 61.944x - 19.442
R² = 0.995
P
y = 45.71x - 14.219
R² = 0.9917
Q
y = 29.822x - 6.7071
R² = 0.9946
Area of Standard / Area of IS
Amount of Standard / Amount of IS

Fig. 7 Calibration curves of TMS-derivatized sterols obtained by GCMS. The figure presents internal standard calibrations for TMS derivatives of: (**a**) pregnenolone; (**b**) coprostanol; (**c**) dehydroergosterol; (**d**) allocholesterol; (**e**) cholesterol; (**f**) 5β-cholestanone; (**g**) dihydrocholesterol; (**h**) episterol; (**i**) cholesta-4,6-dien-3-ol; (**j**) 7-dehydrocholesterol; (**k**) desmosterol; (**l**) zymosterol; (**m**) lathosterol; (**n**) erogosterol; (**o**) cholestenone; (**p**) lanosterol; (**q**) sitosterol. The amount of sample used for derivatization range from 1 to 200 μg, each containing 25 μg of internal standard. Dry samples are derivatized with 100 μL BSTFA/TMCS, and 50 μL *n*-hexane is added, of which 1 μL is injected into the GCMS. Ergosterol can be used as the internal standard for all sterols. For ergosterol, cholesterol is used as the internal standard

4 Notes

Few important precautions during sample preparation include the following:

1. Addition of at least one internal standard per lipid class being analyzed, prior to lipid extraction. It should be noted that the internal standard either be absent in the sample or present in non-significant amounts.
2. Washing of lipid phases using 0.9% NaCl or 1 M KCl at various steps to remove non-lipid contaminants [15].
3. Addition of 0.01% BHT to the solvents. BHT inhibits the enzymatic hydrolysis of lipids and is highly recommended for samples that contain poly-unsaturated lipids [26].
4. Extraction of lipids must be completed as soon as possible once the cells are homogenized. Cell breaking releases active phospholipases which can easily hydrolyze lipids.
5. All extractions must be performed using MS grade solvents and reagents in glass tubes with Teflon-coated screw caps.
6. Lipids extracts must be flushed with N_2 prior to storage.
7. One or more parameters like OD cells used for extraction, dry weight of cells, dry weight of lipids, inorganic phosphate content, and protein content must be recorded, as these are required during normalization of mass spectral signals.

Acknowledgments

We thank grants to AS from ICMR No. 52/08/2019-BMS and University of Lucknow, Lucknow. NB thanks financial support from ICMR No. 56/2/Hae/BMS and his institution.

Financial and Competing Interest Disclosure: *Author has no financial and competing interests with the subject matter or materials discussed in the manuscript.*

Contribution to the Manuscript: *AS, SAU, KA, and NB wrote the manuscript.*

References

1. Sánchez-Guijo A, Hartmann MF, Wudy SA (2013) Introduction to gas chromatography-mass spectrometry. Methods Mol Biol 1065:27–44
2. Pacot GMM, Lee LM, Chin S-T, Marriott PJ (2016) Introducing students to gas chromatography–mass spectrometry analysis and determination of kerosene components in a complex mixture. J Chem Educ 93 (4):742–746
3. Medeiros PM, Simoneit BR (2007) Gas chromatography coupled to mass spectrometry for analyses of organic compounds and biomarkers

as tracers for geological, environmental, and forensic research. J Sep Sci 30(10):1516–1536

4. Snozek CLH, Langman LJ, Cotten SW (2019) An introduction to drug testing: the expanding role of mass spectrometry. Methods Mol Biol 1872:1–10
5. Wang Y, Shen L, Gong Z, Pan J, Zheng X, Xue J (2019) Analytical methods to analyze pesticides and herbicides. Water Environ Res 91 (10):1009–1024
6. Headley JV, Peru KM, Barrow MP (2016) Advances in mass spectrometric characterization of naphthenic acids fraction compounds in oil sands environmental samples and crude oil—a review. Mass Spectrom Rev 35(2):311–328
7. Adhikari PL, Wong RL, Overton EB (2017) Application of enhanced gas chromatography/triple quadrupole mass spectrometry for monitoring petroleum weathering and forensic source fingerprinting in samples impacted by the Deepwater Horizon oil spill. Chemosphere 184:939–950
8. Wudy SA, Schuler G, Sánchez-Guijo A, Hartmann MF (2018) The art of measuring steroids: principles and practice of current hormonal steroid analysis. J Steroid Biochem Mol Biol 179:88–103
9. Liu Z, Weng R, Feng Y, Li Z, Wang L, Su X, Yu C (2016) Fatty acid profiling of blood cell membranes by gas chromatography with mass spectrometry. J Sep Sci 39(20):3964–3972
10. Goad J, Akihisa T (1997) Analysis of sterols, 1st edn. Springer, Dordrecht. https://doi.org/10.1007/978-94-009-1447-6
11. Dufourc EJ (2008) Sterols and membrane dynamics. J Chem Biol 1(1–4):63–77
12. Piironen V, Lindsay DG, Miettinen TA, Toivo J, Lampi AM (2000) Plant sterols: biosynthesis, biological function and their importance to human nutrition. J Sci Food Agric 80 (7):939–966
13. Pizzoferrato L, Nicoli S, Lintas C (1993) GC-MS characterization and quantification of sterols and cholesterol oxidation products. Chromatographia 35(5–6):269–274
14. Jenner AM, Brown SH (2017) Sterol analysis by quantitative mass spectrometry. Methods Mol Biol 1583:221–239
15. Folch J, Lees M, Sloane Stanley GH (1957) A simple method for the isolation and purification of total lipides from animal tissues. J Biol Chem 226(1):497–509
16. Bligh EG, Dyer WJ (1959) A rapid method of total lipid extraction and purification. Can J Biochem Physiol 37(8):911–917
17. Ejsing CS, Sampaio JL, Surendranath V, Duchoslav E, Ekroos K, Klemm RW, Simons K, Shevchenko A (2009) Global analysis of the yeast lipidome by quantitative shotgun mass spectrometry. Proc Natl Acad Sci USA 106(7):2136–2141
18. Mandala SM, Thornton RA, Frommer BR, Curotto JE, Rozdilsky W, Kurtz MB, Giacobbe RA, Bills GF, Cabello MA, Martín I, Palaez F, Harris GH (1995) The discovery of australifungin, a novel inhibitor of sphinganine N-acyltransferase from *Sporormiella australis*. Producing organism, fermentation, isolation, and biological activity. J Antibiot (Tokyo) 48 (5):349–356
19. Singh A, MacKenzie A, Girnun G, Del Poeta M (2017) Analysis of sphingolipids, sterols, and phospholipids in human pathogenic *Cryptococcus* strains. J Lipid Res 58(10):2017–2036
20. Adams BG, Parks LW (1968) Isolation from yeast of a metabolically active water-soluble form of ergosterol. J Lipid Res 9(1):8–11
21. Nes WD, Zhou W, Ganapathy K, Liu J, Vatsyayan R, Chamala S, Hernandez K, Miranda M (2009) Sterol 24-C-methyltransferase: an enzymatic target for the disruption of ergosterol biosynthesis and homeostasis in *Cryptococcus neoformans*. Arch Biochem Biophys 481(2):210–218
22. Kim JH, Singh A, Del Poeta M, Brown DA, London E (2017) The effect of sterol structure upon clathrin-mediated and clathrin-independent endocytosis. J Cell Sci 130 (16):2682–2695
23. Gutiérrez A, del Río JC (2001) Gas chromatography/mass spectrometry demonstration of steryl glycosides in eucalypt wood, Kraft pulp and process liquids. Rapid Commun Mass Spectrom 15(24):2515–2520
24. Singh A, Mahto KK, Prasad R (2013) Lipidomics and in vitro azole resistance in *Candida albicans*. OMICS 17(2):84–93. https://doi.org/10.1089/omi.2012.0075
25. Chang YC, Khanal Lamichhane A, Garraffo HM, Walter PJ, Leerkes M, Kwon-Chung KJ (2014) Molecular mechanisms of hypoxic responses via unique roles of Ras1, Cdc24 and Ptp3 in a human fungal pathogen *Cryptococcus neoformans*. PLoS Genet 10(4):e1004292. https://doi.org/10.1371/journal.pgen.1004292
26. Shiva S, Enninful R, Roth MR, Tamura P, Jagadish K, Welti R (2018) An efficient modified method for plant leaf lipid extraction results in improved recovery of phosphatidic acid. Plant Methods 14:14

Chapter 7

Quantitation of Sphingolipids in Mammalian Cell Lines by Liquid Chromatography–Mass Spectrometry

Nihal Medatwal and Ujjaini Dasgupta

Abstract

Sphingolipids are a major class of bioactive structural lipids that also play diverse roles in signaling events in different disease conditions like cancer. They form a dynamic metabolic network, and alterations in their metabolism are known to contribute to progression of the disease. Therefore, quantitation of these metabolites from mammalian cells as a disease model system can provide us a vivid picture of the regulatory nodes in the pathway that can be targeted to identify sphingolipid-based biomarkers or targets for future therapies. Herein, we present a simple, sensitive, and robust method to simultaneously quantitate sphingolipids of the early steps of the metabolic pathway by a liquid chromatography–mass spectrometry-based method using mammalian cells as model system.

Keywords Sphingolipid metabolism, Ceramides, Glucosylceramides, Sphingosine-1-phosphate, Ceramide-1-phosphate, Sphingolipids, Liquid chromatography–mass spectrometry

1 Introduction

In the past two decades, sphingolipids have emerged as important regulators of various aspects of cancer initiation, progression, and response to anticancer treatments [1, 2]. Dysregulation of sphingolipid metabolic pathway is one of the crucial contributing factors for tumor cells to escape local primary tumor environment, enter into systemic circulation, and establish distant metastasis [3–5]. Research on sphingolipid metabolism is recently being translated to provide clues on early prognosis of disease and development of targeted therapeutics for increased patient survival [6].

In this direction of research, it is necessary to optimize a methodology for analyzing and quantitating the endogenous levels of these bioactive molecules like sphingosine, ceramides, glucosylceramides, sphingomyelins, sphingosine-1-phophate, and ceramide-1-phosphates. Electrospray ionization tandem mass spectrometry (ESI-MS/MS) has emerged as a gold standard for absolute quantitation of sphingolipids, and studies have shown and

Rajendra Prasad and Ashutosh Singh (eds.), *Analysis of Membrane Lipids*, Springer Protocols Handbooks, https://doi.org/10.1007/978-1-0716-0631-5_7, © Springer Science+Business Media, LLC, part of Springer Nature 2020

validated extraction, liquid chromatographic (LC) separation, identification, and quantitation of sphingolipids based on ESI-MS/MS technologies using different mass spectrometers [7–9]. Such studies describing benefits of different mass spectrometric techniques along with a wide array of commercially available natural sphingolipid standards and internal standard cocktails give us a wide perspective to choose the best option and establish and validate a comprehensive method to extract, identify, and quantitate a large number of structural as well as signaling sphingolipids in various model systems.

Literature supports the hypothesis that cancer cell lines have a unique sphingolipid profile that can represent a bio-signature for various cancer types [10, 11]. Therefore, herein, we present a robust method for extraction, liquid chromatographic separation, and ESI-MS/MS-based absolute quantitation of sphingolipids, especially from the early steps of the sphingolipid metabolic pathway, from cancer cells that will provide an invaluable tool to identify deregulations in the sphingolipid pathway.

2 Materials

MCF7 cells obtained from American Type Culture Collection (ATCC, Manassas, VA USA). DMEM media (#D5648, Sigma-Aldrich), Fetal Bovine Serum (FBS) (#10270, Gibco, USA), Dulbecco's Phosphate Buffer Saline (DPBS, #D5652, Sigma-Aldrich), 100× penicillin and streptavidin (#A001A, HiMedia, India) are used for cell culture work. Cell scrappers (#3010) are acquired from Corning (Mexico). Bovine Serum Albumin (BSA) (#500-0206, Bio-Rad, USA) and Bicinchoninic acid (BCA) protein estimation kit (#23227, Thermo Scientific, USA) are used for protein estimation in cell homogenates. The LIPID MAPS internal standard cocktail, sphingolipid mix II (#LM 6005) composed of sphingosine (d17:1), sphinganine (d17:0), sphingosine-1-phosphate (d17:1), sphinganine-1-phosphate (d17:0), ceramide (d18:1/C12:0), glucosylceramide (d18:1/C12:0), lactosylceramide (d18:1/C12:0), sphingomyelin (d18:1/C12:0), and other chain-length subspecies of sphingolipid standards like ceramide (d18:1/C16:0), glucosylceramide (d18:1/C16:0), sphingomyelin (d18:1/C16:0), lactosylceramide (d18:1/C16:0), ceramide-1-phosphate (d18:1/C16:0), and sphingosine-1-phosphate (d18:1) are acquired from Avanti Polar Lipids (Alabaster, USA). LC-MS grade solvents like chloroform (#25669), methanol (#34699), water (#39253), and formic acid (#56302-1L-GL) are obtained from Honeywell (Germany). Potassium hydroxide (# 84749, Sisco Research, India), glacial acetic acid (#CH7C670532, Merck), and Teflon-lined borosilicate tubes (#9900010 from Borosil, India) are used. Glass autoinjector vials (#60180-509) used for loading the samples on UHPLC are obtained from Thermo Fisher Scientific.

3 Methods

3.1 Cell Growth Conditions and Pellet Collection

Breast cancer cells (MCF7) are cultured in 90 mm dishes using DMEM media supplemented with 10% FBS and 1× penicillin and streptavidin. Cells are grown in 37 °C incubator with 5% CO_2 and 95% humidity. Confluent cells (80%) are rinsed twice with DPBS to remove excess media, scrapped from the dish, and collected as cell pellet. Cell pellet (*see* **Note 1**) samples are stored in −80 °C. Protein estimation is done from cell extracts, and equivalent estimate of different cell extracts are taken for lipid isolation.

3.2 Extraction of Sphingolipids from Cell Pellets

Cells pellets are homogenized by probe sonicator (Sonics Vibra Cell VCX130, USA) in DPBS (300 μL) at 30 amp (10 s on and 30 s off) for two cycles. An aliquot (20 μL) is used for protein estimation by bicinchoninic acid (BCA) protein estimation kit. Lipid isolation from each sample (1 mg protein equivalent of each sample) is done using a previously published protocol with modifications [12, 13]. Cell suspensions are transferred to Teflon-lined borosilicate tubes. Chloroform ($CHCl_3$):methanol (MeOH) (1:2, v/v) (3 mL) and LIPID MAPS™ internal standard cocktail sphingolipid mix II (5 μL of 25 μM/125 pmol of each lipid species) are added to the suspension. We can scale down all samples from the same batch to perform lipid isolation and reduce the internal standard amount accordingly if cells samples are less than 1 mg protein equivalent. For example, 0.5 μL of internal standard (from 25 μM stock solution) should be added for 100 μg protein equivalent sample. The homogenate are dispersed further by sonication (3 × 20 s on and 30 s off cycle, amplitude 30). For optimal extraction of sphingolipids, this single-phase mixture is incubated overnight at 48 °C in a water bath. On cooling, 75 μL of 1 M potassium hydroxide (KOH) dissolved in MeOH is added, sonicated briefly, and incubated in a shaking water bath for 2 h at 37 °C as it helps in cleavage of potentially interfering glycerolipids. After cooling at room temperature, samples are neutralized using glacial acetic acid (3–5 μL). Neutralization of the extract is confirmed by checking a drop on pH paper. One aliquot (1.5 mL) is centrifuged to remove the debris and supernatant is transferred to a new tube as single phase extract for estimation of sphingolipids. For better yield, the remaining insoluble debris can be re-extracted with 1 mL of MeOH:$CHCl_3$ (1:2, v/v), centrifuged and the supernatant added to the original tube having single phase extract. Next, the other aliquot is diluted with $CHCl_3$ (1 mL) and water (3.5 mL), vortexed, centrifuged and lower layer of organic phase is transferred to a new tube (Organic phase extract). The upper part is re-extracted two more times with $CHCl_3$ (1 mL) and each time the lower, organic layer is transferred and combined with rest of organic phase extract (*see* **Note 2**). Solvents from single phase and organic phase extracts are evaporated under N_2 gas (Fig. 1).

The dried samples from single phase extract are reconstituted in mobile phase solvent (300 μL), which is a mixture of solvent A and solvent B in 60:40 ratio, where solvent A is a mixture of CH_3OH, water, and HCOOH in 58:41:1 ratio with 5 mM ammonium formate and solvent B is a mixture of CH_3OH and HCOOH in 99:1 ratio with 5 mM ammonium formate. For organic phase sphingolipids, mixture of solvent A and solvent B in 20:80 ratio (300 μL) is used, where solvent A is mixture of water and HCOOH in 99.8:0.2 ratio with 5 mM ammonium formate and solvent B is mixture of CH_3OH and HCOOH in 99.8:0.2 ratio with 5 mM ammonium formate. The resulting resuspensions from organic and single phase extracts are sonicated for 15 s, vortexed, and centrifuged at 15,871 × g for 5 min. The clear supernatant is then transferred to the autoinjector vial for LC-MS/MS analysis. Ceramide-1-phosphates and sphingosine-1-phosphate are analyzed from single phase extracts, whereas ceramides, glucosylceramides, and lactosylceramides are analyzed from organic phase extracts.

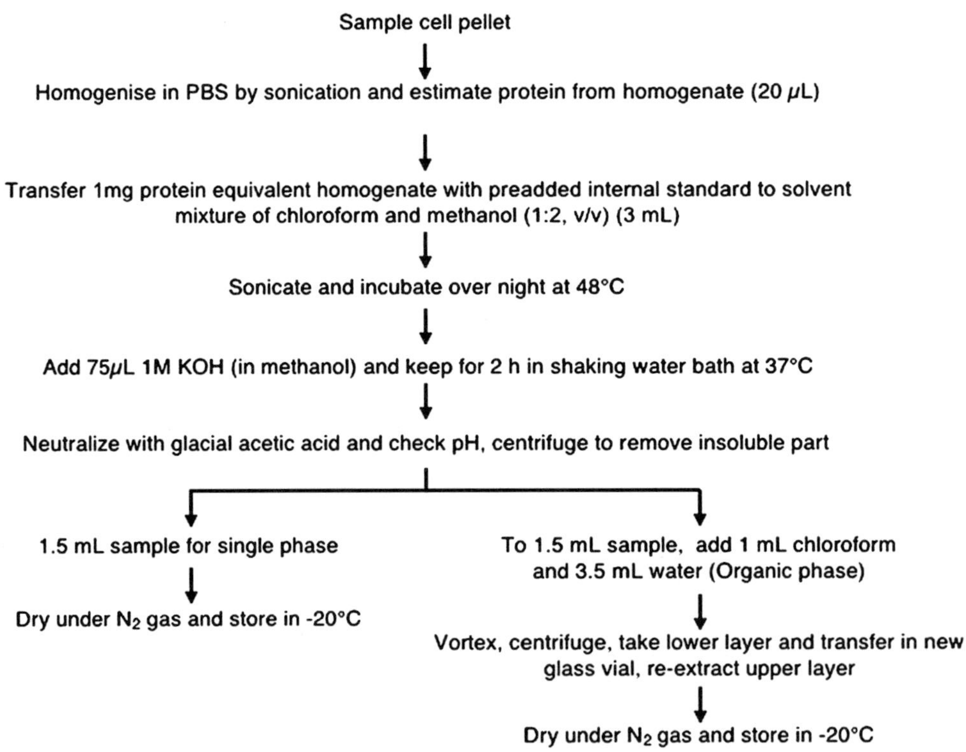

Fig. 1 Workflow diagram of sample preparation to analyze different sphingolipids by LC-MS/MS

3.3 Liquid Chromatographic Separation Method for Different Classes of Sphingolipids

Different classes of sphingolipids are analyzed by reverse phase high pressure UHPLC liquid chromatography (ExionLC AC, SCIEX, USA) using an Acquity BEH C18, 2.1 × 50 mm column (Waters Ltd.) (particle size of 1.7 μm) kept at 60 °C using modified previously published protocol [10, 11]. Total optimized run time for organic phase samples is 16 min and 8 min for single phase samples where solvent A and solvent B are used as mobile phase A and B, respectively, with a total flow rate of 0.3 mL/min. The LC retention time is noted for each analyte. The sample dilution is standardized for each sample by confirming the peak profile and intensity for each transition at that retention time. The analyte peak area at the retention time in all subsequent experiments are integrated in MultiQuant 3.0.2 for data analysis. Representative elution profiles are shown in Fig. 2a, b.

Single phase extracts are used to quantity sphingosine, sphingosine-1-phosphate, and ceramide-1-phosphates as their recovery into the organic phase extract is poor. Solvents A and B are maintained at different ratios for separation of each species with a total run time of 8 min. Solvent A (CH_3OH, water, and HCOOH in 58:41:1 ratio in 5 mM ammonium formate) and solvent B (CH_3OH and HCOOH in 99:1 ratio with 5 mM ammonium formate) are maintained in 60:40 ratio for 0.5 min, followed by a linear gradient of only solvent B (100%) from 0.5 to 7 min, and a wash with mixture of solvent A and B in 60:40 ratio is used from 7 to 8 min before subsequent run.

For organic phase samples, total run time is 16 min to separate multiple species of ceramides, glucosylceramides, lactosylceramides, and sphingomyelins. Solvent A (water/HCOOH in 99.8:0.2 ratio with 5 mM ammonium formate) and solvent B (CH_3OH/HCOOH in 99.8:0.2 ratio with 5 mM ammonium formate) is maintained at 20:80 ratio for 0–6 min, followed only solvent B (100%) for 6–12 min, and finally a wash with mixture of solvent A and B in 20:80 ratio from 12 to 16 min before the next run.

3.4 Mass Spectrometric Analysis of Standards and Cell Samples

3.4.1 Preparation of Standards and Sample/Cell Extract for LC-MS/MS

Mass spectrometric analysis is done using ExionLC AC system coupled to a hybrid triple quadrupole/linear ion trap mass spectrometer (4500 QTRAP, SCIEX, USA) using a binary solvent system. Each commercially obtained sphingolipid standard is dissolved at a concentration of 100 pmol/μL in MeOH. The working concentration of 4 pmol/μL is done by further dilution, using solvent A and B where solvent A (CH_3OH, water and HCOOH in 58:41:1 ratio with 5 mM ammonium formate) and solvent B (CH_3OH and HCOOH in 99:1 ratio with 5 mM ammonium formate) in 60:40 ratio is used for single phase sphingolipids. Solvent A (water and HCOOH in 99.8:0.2 ratio with 5 mM ammonium formate) and solvent B (CH_3OH and HCOOH in 99.8:0.2 ratio with 5 mM ammonium formate) in 20:80 ratio are

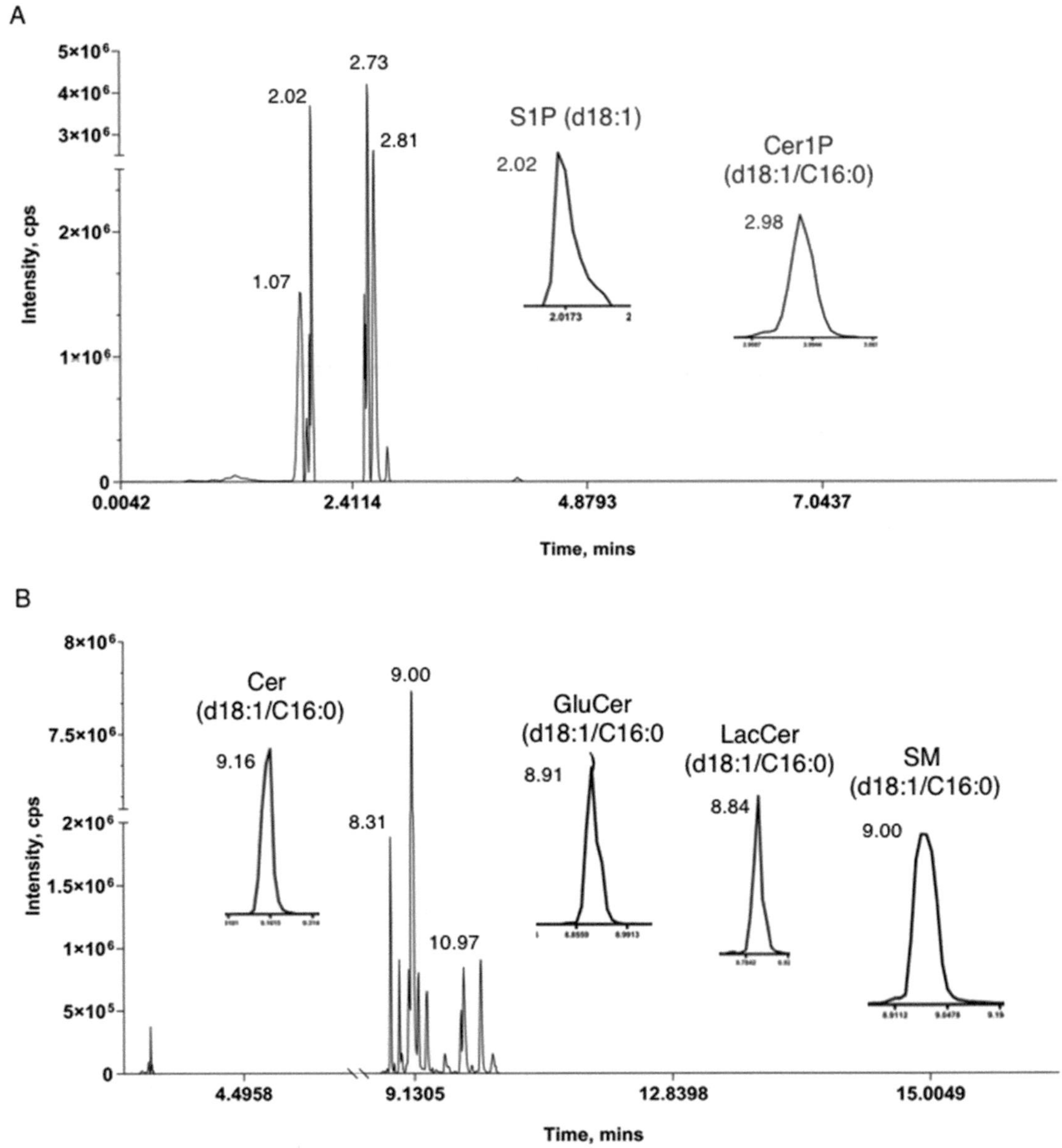

Fig. 2 LC ESI-MS/MS elution profiles of different sphingolipids from single phase (**a**) and organic phase (**b**). (**a**) Elution profiles of ceramide-1-phosphate (C1P) and sphingosine-1-phosphate (S1P) in single phase extract of 1 mg protein equivalent cells with internal standard cocktail. (**b**) Elution profiles of ceramide (Cer), glucosylceramide (GluCer), sphingomyelin (SM), and lactosylceramide (LacCer) from organic phase extract of 1 mg protein equivalent cells with internal standard cocktail, using reverse phase chromatography and analyzed by SMRM in positive ion mode

used as standards for organic phase sphingolipids. Each standard is then serially diluted in cell extract (matrix) to provide the required concentrations for preparing the standard curve (*see* **Note 3**). Matrix dilution is the same for diluting that used for analyte quantitation in cell extract (without standard) (*see* **Notes 4** and **5**). The standards are run in increasing order of concentrations to make the standard curve. This helps to reduce any effect due to carryover. The exact dilution of the sample/cell extract in the appropriate solvent mixture depends on the optimal signal intensity and profile of the peak of the product ion of interest. Standard curve of each representative sphingolipid subspecies is made using the ratio of peak area of analyte and respective internal standard.

3.4.2 Identification of Ionization and Fragmentation Parameters for Standards and Sphingolipid Analytes

Different concentrations of each standard/analyte are infused (5 μL/min) into the ion source to optimize the sphingolipid ionization conditions and fragmentation pattern. After optimizing the declustering potential (DP) and entrance potential (EP), the Q1 (precursor ion mass) settings are fixed and product ion spectra are determined for a range of collision energies (CE) and the relevant product ions are identified. CE and collision cell exit potentials (CXP) are standardized to achieve the most optimal signal for the product ion as shown in Tables 1 and 2. For sphingosine (d18:1), abundant product ion is doubly dehydrated to form fragment ions of *m/z* 264.4 (Fig. 3a). Sphingosine-1-phosphate (d18:1) is cleaved at the phosphate group and dehydration at 3′-position to yield fragment ion of *m/z* 264.4 for sphingoid (d18:1) base (Fig. 3b). Ceramide fragmentation forms product ions of *m/z* 264.4 for d18:1 species by cleavage of the head group at 1-′-position, dehydration at 3′-position, and cleavage of *N*-acyl chain (Fig. 3c). For glucosylceramide pattern, fragmentation is via cleavage of glucose from 1′-position, dehydration at 3′-position, and cleavage of *N*-acyl chain to product ion with *m/z* 264.4 for d18:1 species (Fig. 3d). Fragmentation of lactosylceramide is by cleavage of the lactose from the 1′-position, dehydration at the 3′-position, and cleavage of the *N*-acyl chain to the product ion at *m/z* 264.4 for d18:1 species (Fig. 3e). For sphingomyelin, cleavage is via phosphocholine headgroup to give most abundant product ion with *m/z* 184.4 (Fig. 3f). Ceramide-1-phosphate raise product ion with *m/z* 264.1 for d18:1 species (Fig. 3g). All fragmentations are done in positive ion mode.

3.4.3 Optimization of Multiple Reaction Monitoring Parameters for Subspecies of Each Sphingolipid Class

On the basis of characteristic fragmentation patterns and maximum intensity of the structure-specific ions obtained for every analyte (representative sphingolipid), the MRM set parameters are determined for the precursor/product ion pairs. Scans for the sample/cell extracts over a wide range, for precursors of *m/z* 264.4 (d 18:1 ceramide) and 184.4 (d 18:1 sphingomyelin) covering short (C14:0) to very long (C24:0) acyl chain subspecies are performed.

The results from this survey are used to develop the MRM protocol for analysis of the sphingolipids to set the Q1 and Q3 (product ion mass) so that most abundant structure specific precursor/product pairs are allowed to pass for quantitation. For

Table 1
Mass spectrometer settings and retention times for reverse chromatographic separation under the described conditions for long chain bases and long chain base phosphates from single phase extract

S. no.	Q1 mass (Da)[a]	Q3 mass (Da)[b]	RT[c]	ID[d]	DP (volts)[e]	CE (volts)[f]	CXP (volts)[g]
1	590.400	264.400	2.43	Ceramide-1 phosphate (d18:1/C14:0)	81.00	43.00	12.00
2	618.700	264.400	2.98	Ceramide-1-phosphate (d18:1/C16:0)	94.00	34.00	12.00
3	646.500	264.400	3.17	Ceramide-1-phosphate (d18:1/C18:0)	89.00	35.00	12.00
4	674.500	264.400	3.55	Ceramide-1-phosphate (d18:1/C20:0)	89.00	35.00	12.00
5	702.500	264.400	3.71	Ceramide-1-phosphate (d18:1/C22:0)	89.00	35.00	12.00
6	730.900	264.400	4.25	Ceramide-1-phosphate (d18:1/C24:0)	89.00	36.76	12.00
7	728.400	264.400	3.96	Ceramide-1-phosphate (d18:1/C24:1)	89.00	35.00	12.00
9	380.400	264.400	2.02	Sphingosine-1-phosphate (d18:1)	70.00	23.00	12.00
11	300.400	264.400	1.94	Sphingosine (d18:1)	70.00	25.00	12.00
13	286.500	268.400	1.75	Sphingosine (d17:1)	60.00	17.00	16.00
14	366.302	250.300	1.72	Sphingosine-1-phosphate (d17:1)	60.00	21.00	18.00
19	562.500	264.400	2.56	Ceramide-1-phosphate (d18:1/C12:0)	86.00	41.00	20.00

[a]Q1: Precursor ion mass
[b]Q3: Product ion mass
[c]RT: Retention time
[d]ID: Identified lipid
[e]DP: Declustering potential
[f]CE: Collision energy
[g]CXP: Collision cell exit potential (CXP)

example, Q1/Q3 (*m/z*) of 538.7/264.4 for ceramide (d18:1/C16:0), 566.5/264.4 for ceramide (d18:1/C18:0), 594.6/264.4 for ceramide (d18:1/C20:0), 622.7/264.4 for ceramide (d18:1/C22:0), 648.7/264.4 for ceramide (d18:1/C24:1), 650.7/264.4 for ceramide (d18:1/C24:0) are used. Optimized Q1, Q3, DP, EP, CE, and CXP parameters for all sphingolipids species are mentioned in Tables 1 and 2.

The MRM settings are now used to run the specific precursor/product ion pairs in positive ion mode using scheduled multiple

Table 2
Mass spectrometer settings and retention times for reverse chromatographic separation under the described conditions for complex sphingolipids from organic phase extract

S. no.	Q1 mass (Da)[a]	Q3 mass (Da)[b]	RT[c]	ID[d]	DP (volts)[e]	CE (volts)[f]	CXP (volts)[g]
1	647.700	184.400	8.23	Sphingomyelin (d18:1/C12:0)	90.000	32.000	12.000
2	482.602	264.400	8.35	Ceramide (d18:1/C12:0)	90.000	30.000	12.000
3	644.700	264.400	8.21	Glucosylceramide (d18:1/C12:0)	90.000	43.000	12.000
4	806.700	264.400	8.14	Lactosylceramide (d18:1/C12:0)	90.000	45.000	12.000
5	700.700	264.400	8.91	Glucosylceramide (d18:1/C16:0)	90.000	40.000	12.000
6	728.700	264.400	9.24	Glucosylceramide (d18:1/C18:0)	90.000	42.250	12.000
7	756.600	264.400	9.66	Glucosylceramide (d18:1/C20:0)	90.000	44.500	12.000
8	784.600	264.400	10.07	Glucosylceramide (d18:1/C22:0)	90.000	45.500	12.000
9	812.600	264.400	10.59	Glucosylceramide (d18:1/C24:0)	90.000	46.000	12.000
10	810.900	264.400	10.12	Glucosylceramide (d18:1/C24:1)	90.000	46.000	12.000
11	510.400	264.400	8.7	Ceramide (d18:1/C14:0)	90.000	33.500	12.000
12	538.500	264.400	9.16	Ceramide (d18:1/C16:0)	90.000	35.000	12.000
13	566.500	264.400	9.47	Ceramide (d18:1/C18:0)	90.000	36.500	12.000
14	594.500	264.400	9.92	Ceramide (d18:1/C20:0)	90.000	38.000	12.000
15	622.400	264.400	10.45	Ceramide (d18:1/C22:0)	90.000	39.500	12.000
16	650.600	264.400	11.03	Ceramide (d18:1/C24:0)	90.000	41.000	12.000
17	648.900	264.400	10.49	Ceramide (d18:1/C24:1)	90.000	41.000	12.000
18	675.500	184.400	8.55	Sphingomyelin (d18:1/C14:0)	90.000	33.500	12.000
19	703.500	184.400	9.00	Sphingomyelin (d18:1/C16:0)	90.000	35.000	12.000
20	731.600	184.400	9.29	Sphingomyelin (d18:1/C18:0)	90.000	36.500	12.000
21	759.600	184.400	9.71	Sphingomyelin (d18:1/C20:0)	90.000	38.000	12.000
22	787.600	184.400	10.2	Sphingomyelin (d18:1/C22:0)	90.000	39.500	12.000
23	815.900	184.400	10.75	Sphingomyelin (d18:1/C24:0)	90.000	41.000	12.000
24	862.700	264.400	8.84	Lactosylceramide (d18:1/C16:0)	90.000	48.000	12.000
25	974.800	264.400	10.42	Lactosylceramide (d18:1/C24:0)	90.000	65.000	12.000

[a]Q1: Precursor ion mass
[b]Q3: Product ion mass
[c]RT: Retention time
[d]ID: Identified lipid
[e]DP: Declustering potential
[f]CE: Collision energy
[g]CXP: Collision cell exit potential (CXP)

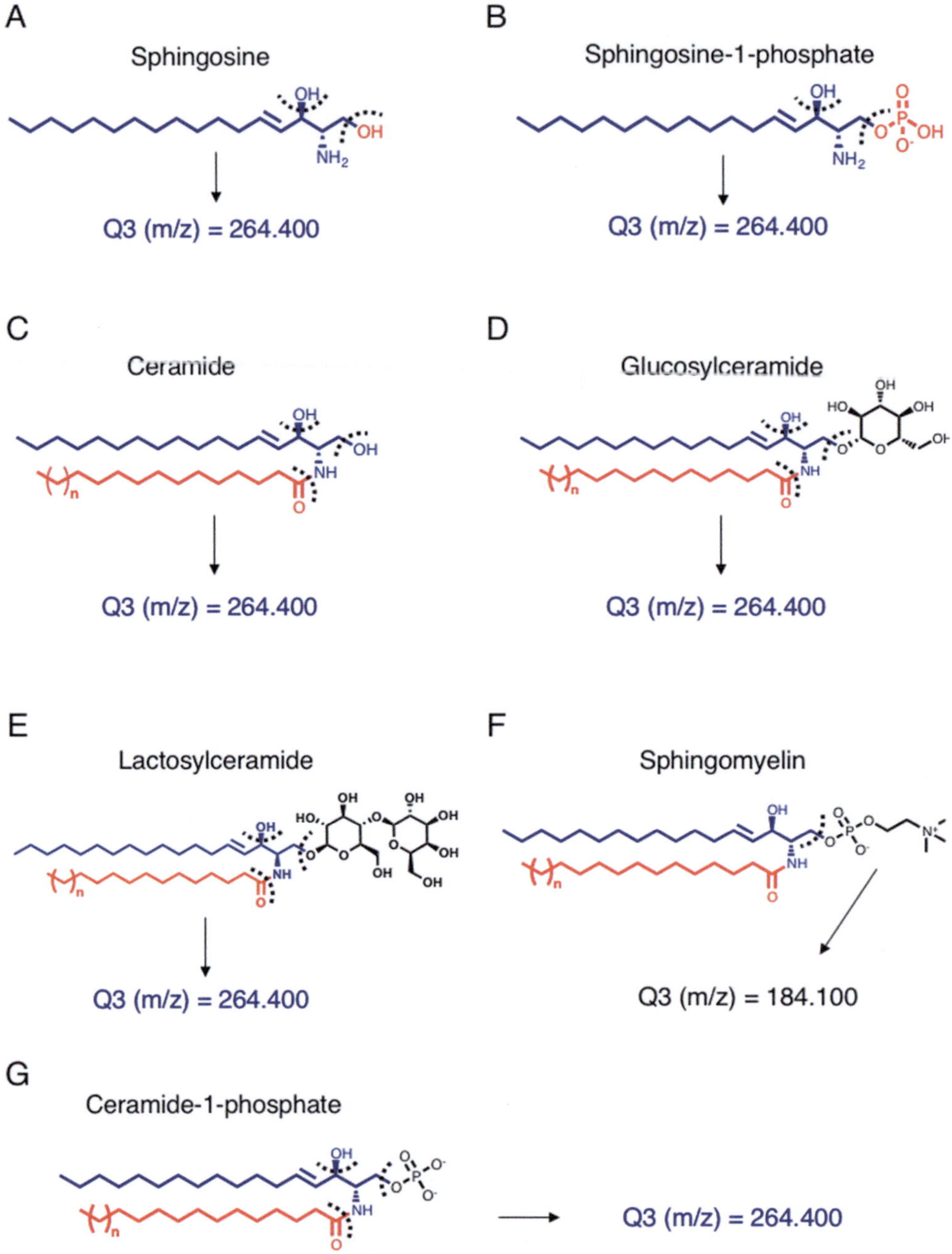

Fig. 3 Mass spectrometry fragmentation summary of different classes of sphingolipids. The sphingoid base and fatty acid are represented in blue and red, respectively. The headgroup is at position 1 and sites of fragmentation are given in dotted lines. Sphingosine, sphingosine-1-phosphate, ceramide, glucosylceramide, lactosylceramide are fragmented into m/z 264.4; sphingomyelin gets fragmented into *m*/*z* 184.4

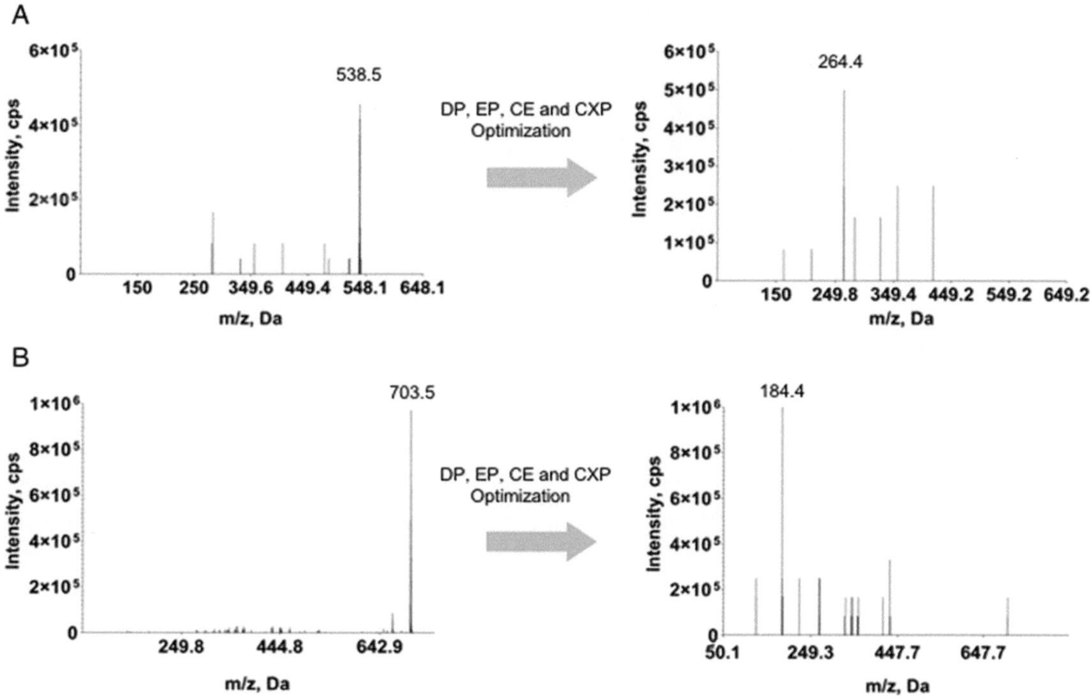

Fig. 4 (**a–b**) Enhanced product ion (EPI) scan for ceramide (Cer) (d18:1/16.0) (**a**) and sphingomylein (SM) (**b**) (d18:1/16.0), before and after optimization of declustering potential (DP), collision energy (CE), entrance potential (EP), and collision cell exit potential (CXP)

reaction monitoring (MRM) with preidentified retention time and elution conditions as described above (Subheading 7.3.3). MRM to enhanced product ion (EPI) scan is used as shown for representative analytes, ceramide (d18:1/C16:0), and sphingomyelin (d18:1/C16:0) in Fig. 4a, b. Source-dependent parameters like nebulizer gases GS1 (50 psi), GS2 (50 psi), curtain gas (25 psi), temperature (600 °C), and ion spray voltage (5.5 kV) are optimized. Four blank runs using same solvent system are given to minimize carryover from sample to sample (*see* **Note 6**).

Extracted ion chromatogram (XIC) for each analyte in a sample cell extract needs to be checked properly to see that internal standard and analyte peak are coming at same retention time. Each sample is run as multiple technical replicates (at least 5) to check for instrument efficiency and percentage coefficient of variation (%CV) should be below 20%. The mass spectrometer acquires the data using software Analyst 1.6.3 (SCIEX, USA) (*see* **Note 7**).

Standard curves are generated for absolute quantitation of sphingolipids. Linearity of the calibration curve is ascertained for ceramide (d18:1/16:0) (0.15–20 pmol), ceramide-1-phosphate (d18:1/16:0) (0.78–20 pmol), lactosylceramide (d18:1/16:0) (0.78–20 pmol), sphingosine-1-phosphate (18:1)

Conc. (pmol)	Ratios
0.15625	0.42
0.3125	0.64
0.625	1.49
1.25	3.03
2.5	6.37
5	15.58
10	25.19
20	47.38

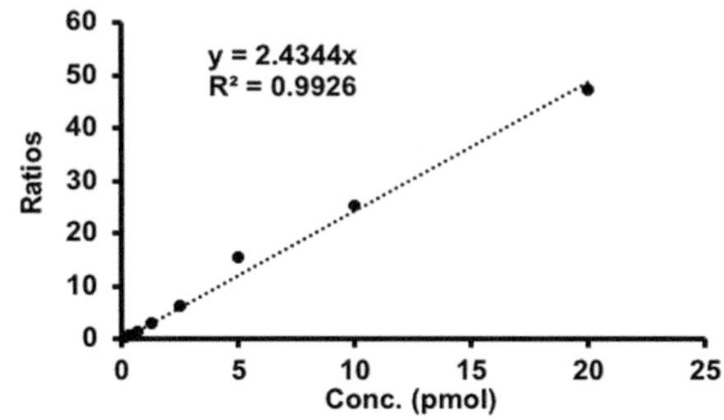

Fig. 5 Standard curve representation for ceramide (Cer) (d18:1/16.0). *X*-axis shows concentration in pmol and *Y*-axis represents ratio of peak area of analyte and internal standard

(0.78–20 pmol), and glucosylceramide (d18:1/16:0) (0.39–50 pmol) using the corresponding standard stock solutions as detailed above (Subheading 7.3.4.1) using 10 μL of injection volume. Linear regression lines are plotted with correlation coefficient values of at least 0.99 (Fig. 5). To quantify absolute value of sphingolipids in samples, the extracted ion chromatograms are examined (*see* **Note 8**).

The area under the peaks for both analytes and spiked sphingolipid internal standard are estimated, and ratio of each analyte with respective internal standard, ceramide (d18:1/12:0), glucosylceramide (d18:1/16:0), lactosylceramide (d18:1/16:0), ceramide-1-phosphate (d18:1/16:0), and sphingosine-1-phosphates (d17:1) are determined. These ratio values are used to quantitate the sphingolipids (pmol/mg protein) in cell samples equivalent to 1 mg protein or any other equivalent.

4 Data Processing and Analysis

To quantify the sphingolipid analytes, the elution profile for each MRM pair should be checked. Data analysis can be divided into two parts, retrieval of raw data and peak integration followed by absolute quantitation of analyte. The .wiff files are transferred to processing computer and opened in MultiQuant 3.0.2. (SCIEX, USA). The peak areas for the analyte, internal standard, and respective residual peaks from blank are integrated by the software using identical integration settings. Noise percentage, baseline sub window, and peak splitting are 40%, 2 min, and 2 points, respectively.

Data processing of samples are done after transferring the data from software to MS Excel. Each integrated analyte peak area is subtracted with respective blanks. Blank subtracted analyte and internal standard peak area ratios can be used for quantitation of sphingolipids in pmol/mg protein. Ratios can be used for making representative data such as in the form bar graphs or heat map (in case multiple sample in one time). However, peak area can be integrated manually to make all sample in one frame and to increase the confidence.

5 Data Interpretation

Data interpretation is the main part of all analysis studies. Data obtained for estimation of sphingolipids from different cell lines can be analyzed with respect to previous studies as well as evaluated for new aspects that emerge from the alterations of lipid flux through the metabolic pathway and deregulations that are reflected. A lot of emphasis should be given on biological replicates for each cell line and consistency of data in multiple technical replicates within each biological replicate (*see* **Notes 9** and **10**).

6 Notes

1. Cell growth conditions are monitored closely and cell growth is avoided under any stress (appearance of black particles and unhealthy look of cells) as this leads to addition of unwanted variables during the experiment. Cell culture conditions should be maintained as consistent as possible.
2. Organic phase extraction: Sample is vortexed till the lower phase became transparent and then it is centrifuged. During recovery of analyte, organic phase is extracted from upper layer at least twice to maximize recovery for each sample.
3. All sphingolipid standard stocks and working solutions are brought to room temperature (when taken out of 20 °C freezer), thoroughly vortexed (so that there is no crystalline lipid precipitate), and centrifuged.
4. Dilution decision: After optimizing all run parameters, different dilutions of each sample (1:2, 1:4, 1:8, 1:16) are run to check the intensity and peak of the MRMs (Multiple Reaction Monitoring) for each dilution of each sample. The common dilution to be used for all samples.
5. As Sphingosine-1-phosphate and Ceramide-1-phosphate are very less in amount in single phase samples so we make dilutions in lower range.
6. Carryover is avoided by running three to four blanks between two samples using the same solvent mixture used for eluting the samples. Column is washed thoroughly after 48–72 h of run with CH_3CN:CH_3OH:water (5:4:1; v/v/v) for 6 h or more.

7. There will be chances of a shift in retention time if solvent/buffer conditions or pH are changed.
8. Lower limit of quantitation and upper limit of quantitation is checked for each analyte during preparation of standard curve to bring it within quantitation range.
9. The method described here cannot differentiate between the two monohexosyl ceramides, i.e. glucosylceramides and galactosylceramides. For samples/cell extracts that have galactosylceramides along with glucosylceramide, they need to be resolved using long, normal phase silica column by isocratic elution with a solvent system (CH_3CN:CH_3OH:HCCOH, 97/2/1 v/v/v with 5 mM ammonium acetate) and specific internal and natural standards [12]. However, galactosylceramides are primarily found in brain and nervous tissue and are very low in other organs. Therefore, since, we have used MCF7 cells (primary breast cancer cell line) in this study, we have referred to the monohexosyl ceramides as glucosylceramides.
10. Information regarding structure, fragmentation patterns, standards, pathways is available in LipidMaps (www.lipidmaps.org) and the affiliated SphinGOMAP (www.sphingomap.org) for ready reference.

Acknowledgments

The support from Amity University Haryana, India is acknowledged. U.D. is supported by BT/PR19624/BIC/101/488/2016 (DBT) and ECR/2016/001603 (DST). Amity Lipidomics Research Facility at Amity University Haryana is supported by DST-FIST grant SR/FST/LSI-664/2016.

References

1. Hannun YA, Obeid LM (2018) Sphingolipids and their metabolism in physiology an disease. Nat Rev Mol Cell Biol 19(3):175–191
2. Ogretmen B (2018) Sphingolipid metabolism in cancer signalling and therapy. Nat Rev Cancer 18:33–50
3. Ryland LK, Fox TE, Liu X, Loughran TP, Kester M (2011) Dysregulation of sphingolipid metabolism in cancer. Cancer Biol Ther 11 (2):138–149
4. Pyne NJ, Pyne S (2010) Sphingosine 1-phosphate and cancer. Nat Rev Cancer 10 (7):489–503
5. Ogretmen B, Hannun YA (2004) Biologically active sphingolipids in cancer pathogenesis and treatment. Nat Rev Cancer 4(8):604–616
6. Shaw J, Costa-Pinheiro P, Patterson L, Drews K, Spiegel S, Kester M (2018) Novel sphingolipid-based cancer therapeutics in the personalized medicine era. Adv Cancer Res 140:327–366
7. Bielawski J, Szulc ZM, Hannun YA, Bielawska A (2006) Simultaneous quantitative analysis of bioactive sphingolipids by high-performance liquid chromatography-tandem mass spectrometry. Methods 39:82–91

8. Scherer M, Leuthäuser-Jaschinski K, Ecker J, Schmitz G, Liebisch G (2010) A rapid and quantitative LC-MS/MS method to profile sphingolipids. J Lipid Res 51:2001–2011
9. Merrill AH Jr, Sullards MC, Allegood JC, Kelly S, Wang E (2005) Sphingolipidomics: high-throughput, structure-specific, and quantitative analysis of sphingolipids by liquid chromatography tandem mass spectrometry. Methods 36:207–224
10. Nunes J, Naymark M, Sauer L, Muhammad A, Keun H, Sturge J, Stebbing J, Waxman J, Pchejetski D (2012) Circulating sphingosine-1-phosphate and erythrocyte sphingosine kinase-1 activity as novel biomarkers for early prostate cancer detection. Br J Cancer 106(5):909–915
11. Pyne S, Edwards J, Ohotski J, Pyne NJ (2012) Sphingosine 1-phosphate receptors and sphingosine kinase 1: novel biomarkers for clinical prognosis in breast, prostate, and hematological cancers. Front Oncol 2:168
12. Shaner RL, Allegood JC, Park H, Wang E, Kelly S, Haynes CA, Sullards MC, Merrill AH Jr (2009) Quantitative analysis of sphingolipids for lipidomics using triple quadrupole and quadrupole linear ion trap mass spectrometers. J Lipid Res 50:1692–1707
13. Dasgupta U, Bamba T, Chiantia S, Karim P, Tayoun AN, Yonamine I, Rawat SS, Rao RP, Nagashima K, Fukusaki E, Puri V, Dolph PJ, Schwille P, Acharya JK, Acharya U (2009) Ceramide kinase regulates phospholipase C and phosphatidylinositol 4, 5, bisphosphate in phototransduction. Proc Natl Acad Sci USA 106:20063–20068

Chapter 8

Exploring Membrane Lipid and Protein Diffusion by FRAP

Parijat Sarkar and Amitabha Chattopadhyay

Abstract

Knowledge of membrane dynamics is crucial since it allows us to understand membrane function. Fluorescence recovery after photobleaching (FRAP) is a widely used technique to monitor diffusion of lipids and proteins in biological membranes. We outline here general aspects of FRAP, followed by a step-by-step guide to carry out FRAP measurements for exploring diffusion of fluorescently labeled lipids and proteins in membranes of attached cells and membranes of *Candida albicans*. In this process, we have provided detailed hands-on tips, judicious use of which would ensure reliability and quality of acquired FRAP data and associated analysis.

Keywords FRAP, Confocal microscopy, Lateral diffusion, Mobile fraction, Diffusion coefficient, DiI, GFP

1 Introduction

Fluorescence recovery after photobleaching (FRAP) is one of the most widespread approaches for quantitative analysis of lateral diffusion characteristics in membranes [1–6]. Although the basic principles of FRAP (*see* [7] for a lucid description of the early history of FRAP) were developed almost four decades ago [8–12], the technique has experienced a resurgence due to the development of photostable fluorescence probes, generation of green fluorescent protein (GFP, see later), and rise of commercially available confocal microscopes [13, 14]. FRAP is a photoperturbation technique that involves generation of a concentration gradient of fluorescent molecules by irreversibly photobleaching (i.e., a photo-induced covalent modification of fluorophores that extinguishes its fluorescence) a fraction of fluorophores in a selected region. By making a fraction of fluorescent molecules invisible, FRAP alters the steady fluorescence intensity in a region of the cell without disrupting or creating any protein gradients. The dissipation of this fluorescence gradient with time occurs as the surrounding unbleached fluorescent molecules re-equilibrate in the bleached

Rajendra Prasad and Ashutosh Singh (eds.), *Analysis of Membrane Lipids*, Springer Protocols Handbooks,
https://doi.org/10.1007/978-1-0716-0631-5_8,

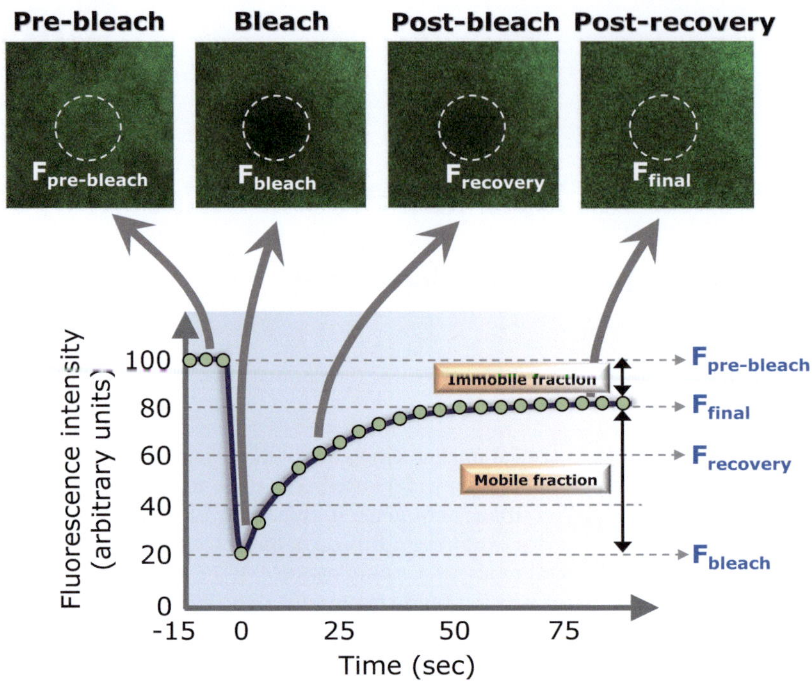

Fig. 1 The basic design of a FRAP measurement. The initial total fluorescence intensity in a region of interest (ROI) on the cell surface before photobleaching is denoted as $F_{\text{pre-bleach}}$. A concentration gradient of fluorescent molecules is generated by irreversibly photobleaching a population of fluorophores in the ROI (shown as a dashed circle) using a high-power laser beam. The total fluorescence intensity in the ROI immediately after photobleaching is shown as F_{bleach}. The concentration gradient of fluorophores gets dissipated with time due to lateral diffusion of unbleached fluorophores (outside ROI) into the bleached region and bleached fluorophores (inside ROI) away from the bleached region. The total fluorescence intensity in the ROI after complete fluorescence recovery is termed as F_{final}. Analysis of the rate of recovery of fluorescence (from F_{bleach} to F_{final}) yields the lateral diffusion coefficient (D). The extent of fluorescence recovery provides information on the fraction of molecules that are mobile in this time scale (termed mobile fraction, M_f)

region, which is then monitored (*see* Fig. 1). The extent and rate at which fluorescence recovery takes place can be quantified to describe the diffusion parameters. Two key diffusion parameters of a molecule can be obtained from quantitative FRAP measurements: the mobile fraction (M_f), which is the fraction of molecules that can diffuse into the bleach region during the time course of the measurement, and the diffusion coefficient (D) that reflects the rate of molecular movement. Analysis of FRAP measurements therefore provides information on the diffusion characteristics of an ensemble of molecules, as the area monitored is large and typically in the order of micrometers [15].

Lateral diffusion in biomembranes is a fundamental biophysical process that regulates the dynamics of lipid–protein and protein–protein interactions at the cell surface [16]. Lateral diffusion of receptors in the membrane represents an important determinant of

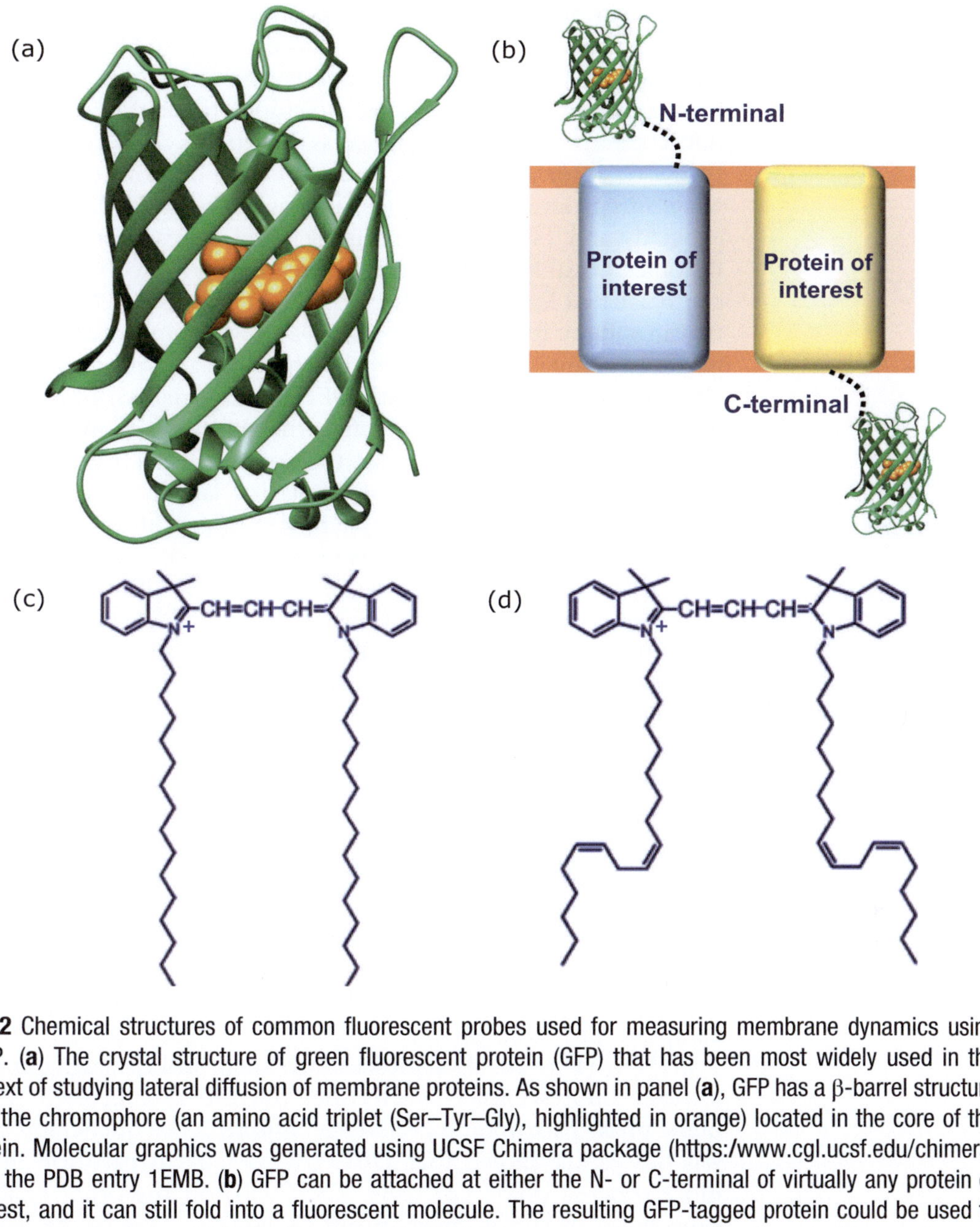

Fig. 2 Chemical structures of common fluorescent probes used for measuring membrane dynamics using FRAP. (**a**) The crystal structure of green fluorescent protein (GFP) that has been most widely used in the context of studying lateral diffusion of membrane proteins. As shown in panel (**a**), GFP has a β-barrel structure with the chromophore (an amino acid triplet (Ser–Tyr–Gly), highlighted in orange) located in the core of the protein. Molecular graphics was generated using UCSF Chimera package (https:/www.cgl.ucsf.edu/chimera) from the PDB entry 1EMB. (**b**) GFP can be attached at either the N- or C-terminal of virtually any protein of interest, and it can still fold into a fluorescent molecule. The resulting GFP-tagged protein could be used to study protein dynamics. (**c**) $DiIC_{18}(3)$ and (**d**) *FAST* DiI are common fluorescent probes used to quantify lipid dynamics in membranes

the overall efficiency of the signal transduction process [17]. In this context, GFP and its variants have become popular reporter molecules for monitoring expression, localization, and mobility of various membrane proteins by tagging it to the N- or C-terminal of the protein of interest [18–22] (Fig. 2). The use of GFP-tagged proteins to study membrane dynamics has a number of advantages:

(1) cellular transcription and translation ensure the presence of receptors covalently attached to fluorescent proteins in cells and eliminate the necessity of labeling receptors with fluorescent ligands before each experiment, (2) the stoichiometry of the receptor and fluorescent protein is well defined as the latter is covalently attached to the receptor at the DNA level, (3) complications encountered while using fluorescent ligands such as dissociation of ligands are avoided, and (4) analysis of unliganded states of the receptor becomes possible, which is otherwise not permitted with fluorescently labeled ligand. Another remarkable aspect of GFP fusion is that in spite of its large size (~27 kDa), most (but not all, *see* [23]) proteins maintain their native biochemical and pharmacological characteristics after fusion with GFP (or its variants). Importantly, the physical principles that define the diffusion of molecules in membranes are different from that of molecules diffusing in a bulk solvent. This is due to the fact that lateral diffusion in the membrane is relatively insensitive to size of the diffusing molecule since the diffusion coefficient is proportional to the logarithm of the reciprocal of the hydrodynamic radius of the diffusing molecule [22, 24]. As a consequence, unlike diffusion of soluble proteins in the cytoplasm, the size dependence of the diffusion of membrane proteins is rather subtle, and in spite of relatively large size of the GFP tag (~27 kDa), its effect on diffusion parameters is minimal.

Fluorescently labeled lipid analogs are widely used for measuring lateral dynamics of lipids in membranes [25, 26]. The DiI group of lipids are well characterized and commonly used probes for such measurements. They are composed of a polar indocarbocyanine headgroup and two hydrophobic alkyl chains (*see* Fig. 2c, d) that impart an overall amphiphilic character. These probes have earlier been shown to preferentially partition into fluid (disordered) or gel (ordered) phases of the membrane depending on the degree of similarity between their acyl chain length and those of lipids that comprise the host plasma membrane [27–29]. $DiIC_{18}(3)$ and *FAST* DiI (Fig. 2c, d) represent two such probes having similar intrinsic fluorescence characteristics but differing in their membrane phase partitioning preference. Fluorescence quenching analysis has earlier shown that $DiIC_{18}(3)$ prefers to partition into a more ordered phase [27, 28]. *FAST* DiI, on the other hand, is expected to partition more into disordered regions of the plasma membrane due to unsaturation in its acyl chains that would introduce kinks in the acyl chain leading to packing defects in the membrane [26, 30]. This is further supported by the observed similarities in endosomal trafficking properties of *FAST* DiI with short-chain DiI analogs [31] that are known to preferentially partition into a more disordered phase of the membrane [28]. Importantly, we previously showed that $DiIC_{16}(3)$ (a probe similar to $DiIC_{18}(3)$) displays a significant extent of detergent insolubility (a property of ordered membrane domains) relative to *FAST* DiI in cell membranes, which possibly could reflect partitioning preferences of these probes into different membrane domains [29]. We previously analyzed lateral diffusion characteristics of DiI group of fluorescent lipids in natural membranes using FRAP [32]. In another study, *FAST* DiI was used to monitor lipid diffusion in membranes of the wild type and mutants lacking ergosterol of the pathogenic yeast *Candida albicans* [33]. Interestingly, by measuring lipid diffusion in *C. albicans*, we demonstrated that lipid dynamics in membranes of the wild type and ergosterol biosynthetic mutants of *C. albicans* correlates well with their drug resistance characteristics.

In this protocol, we focus on FRAP measurements to assess plasma membrane dynamics utilizing fluorescent lipid analogs and fluorescently labeled proteins. We describe here basic principles of FRAP measurements, time-lapse imaging, bleaching, and recovery of fluorescence with an emphasis on compulsory controls and finally discuss some aspects of data processing.

2 Materials

(a) Solvents and buffers.
   - (i) Freshly prepared 1 M sorbitol, 0.1 M EDTA buffer (buffer A).
   - (ii) Spectroscopy grade methanol and ethanol.
   - (iii) Phosphate buffer saline (PBS; pH 7.4).

(b) Reagents.
   - (i) Stock solution of $DiIC_{18}(3)$ and *FAST* DiI in spectroscopy grade methanol (*see* **Note 1**).
     - Measure the stock concentration using the molar extinction coefficient of 148,000 M^{-1} cm^{-1} at 549 nm [32].
   - (ii) Poly-L-lysine (0.01% w/v) in Milli-Q water.
   - (iii) Glass coverslip (22 mm); thickness 0.17 mm or commonly known as #1.5 coverslips (*see* **Note 2**).

3 Methods

(a) *Labeling of adherent cells.*
   - (i) Grow cells on Lab-Tek chamber slides under the required experimental conditions.
   - (ii) Wash twice in cold PBS before labeling them with fluorescent lipid probes.

(iii) Label the plasma membrane of cells using 15 μM (final concentration) *FAST* DiI for 30 min or 8 μM (final concentration) $DiIC_{18}(3)$ for 60 min at 4 °C.
- Stock solutions of the DiI probes are diluted in PBS to prepare the labeling solutions making sure that the residual methanol concentration is always <0.1% (v/v).
- In case of analysis of lateral diffusion of membrane proteins, skip **Steps (ii)** and **(iii)**, and use a suitable cell line expressing the fluorescently tagged protein of interest (*see* **Note 3**).

(iv) Labeled cells are washed twice in PBS.

(v) FRAP measurements are performed on cells in PBS containing 1 mM $CaCl_2$ and 0.5 mM $MgCl_2$.

(b) *Labeling of yeast/Candida cells.*

(i) Suspend an overnight grown culture of cells at a density of ~10^8 cells/ml in buffer A.

(ii) Label the plasma membrane of cells using 10 μM *FAST* DiI (final concentration) or 8 μM (final concentration) Dil-$C_{18}(3)$ in dark at ~25 °C with mild shaking in glass tubes for 30 min.
- In case of analysis of lateral diffusion of membrane proteins, skip **Step (ii)** and use a suitable strain expressing the fluorescently-tagged protein of interest (*see* **Note 3**).

(iii) Centrifuge and wash twice with buffer A.

(iv) Resuspend the cell pellet in 50 μl of buffer A.

(v) Transfer an aliquot of 10 μl of this suspension to a clean glass slide and mount between a glass slide and a poly-L-lysine pre-coated glass coverslip.

(vi) Seal the ends of the coverslip with nail enamel.

(c) *FRAP measurements.*

(i) FRAP measurements could be performed in an inverted confocal microscope equipped with high numerical aperture objective and suitable laser sources compatible with the fluorophore of interest. The choice of the laser excitation, dichroic mirrors, and emission filters is determined by the fluorophore used for a particular experiment. The choice of the objective lens, zoom factor, and pinhole settings would depend on the size of the cell and the thickness and size of the region of interest (ROI) being monitored. The detector gain could be changed in order to image cells with different protein expression levels or difference in labeling intensities. The detector gain should be set to maximize the signal-to-noise ratio while avoiding saturation in any pixel (*see* **Note 4**).

- $DiIC_{18}(3)$ and *FAST* DiI could be excited using 543 nm line of a helium–neon laser (or 561 nm DPSS laser), and fluorescence emission could be collected using the 565–630 nm band-pass filter.
- GFP could be excited using 488 nm line of an argon laser, and fluorescence emission could be collected using the 500–550 nm band-pass filter.

(ii) Identify the cell of interest on the confocal microscope and bring it to the desired focus. Define a circular region of interest (ROI) of ~1 μm in radius which will be subsequently photobleached and monitored (*see* **Notes 5** and **6**).

(iii) The rule of thumb for any FRAP measurement is to achieve significant fluorescence photobleaching within a very short period of time. For successful photobleaching, high laser power with short bleach time should be chosen over low laser power with long bleach time (*see* **Note 7**). Theoretically, bleaching duration should be infinitesimally small compared to characteristic diffusion time (τ_d, the time required for half-maximal recovery of fluorescence). In reality, one could achieve ~70–80% photobleaching within ~100 ms using any modern confocal microscope.

(iv) Next, the photobleaching parameters (scan speed, laser power, and bleaching iterations) are determined empirically. The number of times (bleaching iterations) the laser scans the ROI to achieve significant extent of photobleaching is an important parameter. This number would depend on intrinsic factors such as mobility and photostability of the fluorophore for a given system, and on extrinsic factors such as laser power and scan speed of the laser beam. It is possible to use a different scan speed (with more number of bleaching iterations) during bleaching in order to achieve significant fluorescence photobleaching within a very short period of time. Typically, fluorescence in the bleach ROI should be reduced by 70–80% of the initial value in order to observe appreciable fluorescence recovery (*see* **Note 8**). Higher extents of photobleaching often induce artifacts in FRAP experiments since it could damage the sample or induce cross-linking of fluorophores (*see* **Note 9**).

(v) It is also necessary to avoid undesired photobleaching of the ROI while imaging (scanning) during pre- and post-bleach time points (*see* **Note 10**). Scanning laser power should be set to the lowest possible power necessary for sufficient signal-to-noise of the fluorescence intensity over background (*see* **Note 10**). Typically, imaging parameters (dye concentration, laser power, and PMT voltage) should be adjusted in a way such that pre-bleach intensity is ~100 times more than background fluorescence.

(vi) Further optimization involves tuning imaging parameters (beam splitter configuration, emission bandpass, and pinhole diameters) to achieve an optimal collection of fluorescence and improving signal-to-noise ratio (*see* **Note 4**).

(vii) The next parameter is the total number of scans (sometimes referred to as cycles) to be acquired during the entire experiment (pre-bleach, bleach, and post-recovery). This number refers to the maximum number of scans before significant photobleaching of the sample becomes apparent (*see* **Note 11**).

(viii) The time interval between successive scans is then determined. This value is dependent on the recovery kinetics of the particular fluorophore being monitored. For molecules undergoing fast diffusion, the time interval would be small (or sometimes zero). In case of slow diffusion, the time interval between successive scans should be large (*see* **Note 12**). To ensure complete recovery of fluorescence after bleaching, it is generally advisable to monitor fluorescence recovery for a period of ~$4\tau_d$.

(ix) After standardizing the number of cycles (*see* points (vii) and (viii) above) to be acquired for the entire experiment, three separate phases of the FRAP measurement are performed. First, images are acquired with low-power laser settings to extract pre-bleach intensities. Typically, five to ten pre-bleach images are acquired before switching to the high-intensity laser pulse for selectively photobleaching the ROI in the shortest possible time. One must make sure that the pre-bleach fluorescence intensity of the sample, as well as the ROI, does not fluctuate significantly. Immediately after photobleaching, the sample is monitored using the same low-power imaging settings (as used for pre-bleach time points) until the required number of cycles is completed.

(x) Laser scanning confocal microscopes offer a number of options for data collection. These include image size (e.g., 512 × 512 or 1024 × 1024), range of data collection (8-bit or 12-bit), and different file formats. It is advised to determine the requirement of the image analysis software prior to data acquisition. Collect 20–30 recovery plots for each treatment condition and repeat the experiment at least on three different days.

(d) *Analysis of FRAP measurements.*

(i) Accurate analysis of FRAP data requires that the bleaching event is much shorter than the recovery time, and the recovery is monitored until a recovery plateau is achieved. After performing FRAP measurements, based on the protocol specified above, data representing the mean fluorescence intensity as a function of time are extracted from the ROI of each time series image (*see* **Note 13**). This data should be then background corrected by subtracting background fluorescence values collected by performing FRAP measurement on an area without cells.

(ii) The background-corrected fluorescence trace is then normalized to both the pre-bleach fluorescence intensity in the ROI and the time of bleach (*see* Fig. 3). The latter is achieved by subtracting the pre-bleach time from each data point on the time axis. This results in the pre-bleach time points starting from $t < 0$, bleach point $t = 0$, and first post-bleach time point starting from a time $t > 0$ (Fig. 3).

(iii) The normalized and background-subtracted fluorescence intensity in the ROI (F(t)) versus normalized time (t) is analyzed according to the uniform-disk illumination model based on theoretical framework given by [34]:

$$F(t) = [F(\infty) - F(0)]\,[\exp(-2\tau_d/t)(I_0(2\tau_d/t) + I_1(2\tau_d/t))] + F(0) \tag{1}$$

where $F(\infty)$ is the post-bleach recovered fluorescence at time $t \to \infty$, $F(0)$ is the bleached fluorescence intensity in the ROI immediately after bleach, and τ_d is the characteristic diffusion time. I_0 and I_1 are modified Bessel functions. We routinely perform nonlinear curve fitting of the recovery data to Eq. (1) using GraphPad Prism software version 4.0 (San Diego, CA).

- For FRAP data fitting using GraphPad Prism in the "equation type" tab, use the following equation:

  Y = (Fi-Fo)∗(exp(−2∗T/X)∗(besseli(0,2∗T/X) + besseli(1,2∗T/X))) + Fo

  where *Y* is fluorescence intensity at time *X*; Fi = final fluorescence intensity; *Fo* = initial fluorescence intensity; and *T* = characteristic diffusion time (τ_d).

- Feed initial values of Fi, Fo, and *T* by looking at the fluorescence trace. For example, when analyzing the fluorescence recovery trace shown in Fig. 1, choose *Fi* = 80, *Fo* = 20, and *T* = 12.

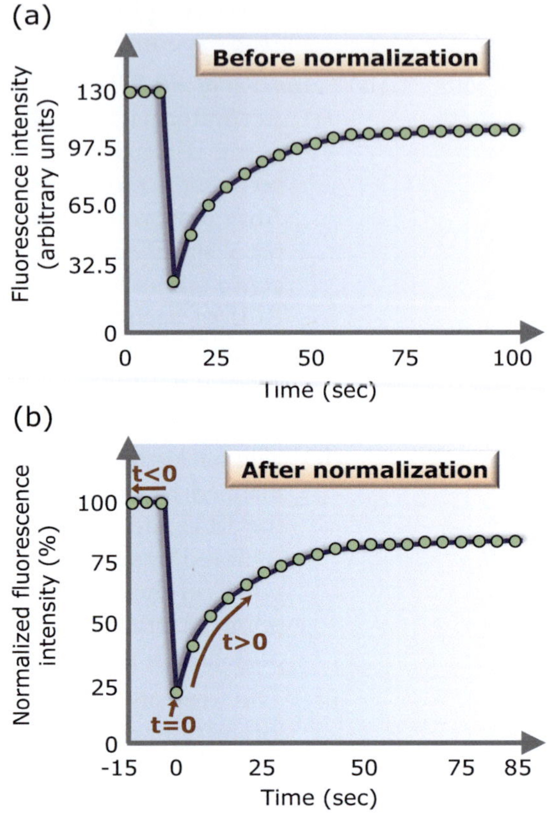

Fig. 3 Normalization of FRAP data. Raw data from FRAP measurements are normalized to both the pre-bleach fluorescence intensity in the ROI and the time of bleach. The latter is achieved by subtracting the total time until the first bleach point from each data point on the time axis. This results in the pre-bleach time points starting from $t < 0$, bleach point $t = 0$, and first post-bleach time point starting from a time $t > 0$

- The equation does not have a solution at $X = 0$. Therefore, remove the data point corresponding to $t = 0$.

(iv) Diffusion coefficient (D) is determined from the following equation:

$$D = \omega^2/4\tau_d \quad (2)$$

where ω is the radius of the ROI (*see* **Notes 14–17**).

(v) M_f is determined from the following equation:

$$M_f = [F(\infty) - F(0)]/[F(p) - F(0)] \quad (3)$$

where $F(p)$ is the mean background corrected and normalized pre-bleach fluorescence intensity, $F(\infty)$ is the post-bleach recovered fluorescence at time $t \rightarrow \infty$, and $F(0)$ is the bleached fluorescence intensity in the ROI immediately after bleach. This ratio results in values between 0 and 1, or when expressed as a percentage, between 0% and 100% (*see* **Note 18**).

4 Notes

1. Light sensitive, store in a brown glass vial at −20 °C.
2. Clean coverslips with 70% ethanol (v/v) and air dry thoroughly.
3. Studying the dynamics of membrane proteins using FRAP requires fusion of a suitable fluorophore to the protein of interest. Desired characteristics for suitable fluorophores include (a) photostability during time-lapse imaging, (b) enough brightness to obtain a high signal-to-noise ratio, (c) absence of photoreversible bleaching, and (d) monomeric nature to avoid trivial association between tagged proteins. In this context, GFP-tagged (or its variants) proteins are ideal for use in FRAP measurements [22]. Additional advantages of GFP-tag include minimal photodamage to the cell during photobleaching [35]. The compact barrel-like structure of GFP shields the external environment from the damaging effects that are caused by the reactive intermediates generated during photobleaching [36–38].
4. It is crucial to adjust the acquisition parameters of the photomultiplier tube (PMT) in order to acquire images in the dynamic range of intensity acquisition. Pixel saturation (pixel intensity that exceeds the detector scale, e.g., >255 for an 8-bit image) during time series imaging would result in inaccurate estimation of the fluorescence recovery profile. This is because saturated pixels only record 255 as the intensity value and the true intensity of the pixel is not registered. The pinhole value, determined by the numerical aperture of the objective lens and the wavelength of the laser, should be kept constant across all images. The pinhole should be opened wide enough to acquire an adequate signal while keeping the laser intensity low to avoid photobleaching during pre- and post-bleach time points (*see* **Note 10**). It should be kept in mind that increasing the pinhole decreases the z resolution of the image by increasing the thickness of the optical section.
5. The bleach spot (ROI) should comprise a relatively small proportion of the cellular pool of fluorescent proteins. One should make sure that the dimension of the ROI is small compared to the size of the cell in order to limit the total loss of fluorescence,

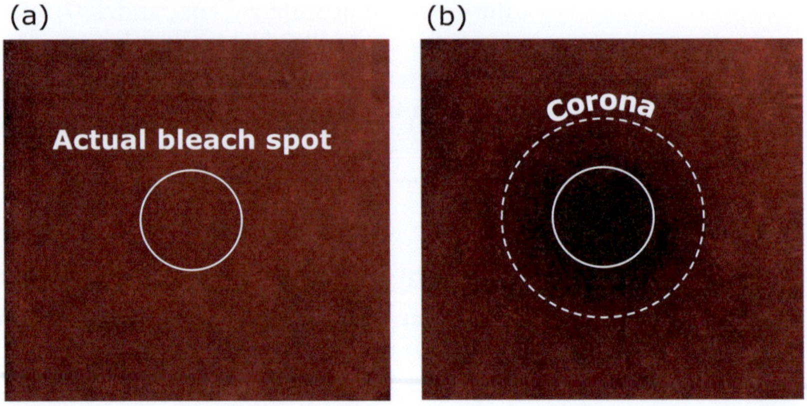

Fig. 4 The formation of a bleached "corona" due to prolonged bleaching in case of rapidly diffusing molecules. The bleach spot broadens with increasing bleaching iterations due to bleached "corona" (dashed circle) that emerges around the specified ROI (solid circle). This "corona" effect leads to a broadening of the bleached region beyond the marked ROI resulting in underestimation of diffusion coefficient

while it should be large enough to get a reasonable signal-to-noise ratio.

6. FRAP measurements are generally performed on the basal surface of cells that are in contact with the glass coverslip. This is because analysis of FRAP data is based on the theoretical framework which assumes that fluorescence recovery into the bleached area is isotropic in the plane of the membrane [34]. This condition is satisfied when one monitors the uniform fluorescent bottom surface of cells attached to the coverslip. In addition, the planar geometry of the uniform bottom surface of cells ensures that the theoretical dimensions of the circular ROI used for photobleaching are not distorted in the actual sample.

7. In case of a rapidly diffusing molecule, photobleaching cannot be assumed to be instantaneous. Since lateral diffusion does not "stop" while an ROI is photobleached, if photobleaching is slow relative to the diffusion of the molecule of interest, then the unbleached molecules will enter the ROI from the edges and become bleached. A longer bleach duration (of the same order of τ_d) would, therefore, result in the formation of a "corona" around the bleach spot [39] (*see* Fig. 4). This results in an effective increase in bleach spot radius and subsequently underestimates the value of diffusion coefficient. Experiments should be performed under conditions where the laser power is set to its maximum (100%) to achieve the shortest possible bleach period. For correction of the "corona" effect, please refer to **Note 14**.

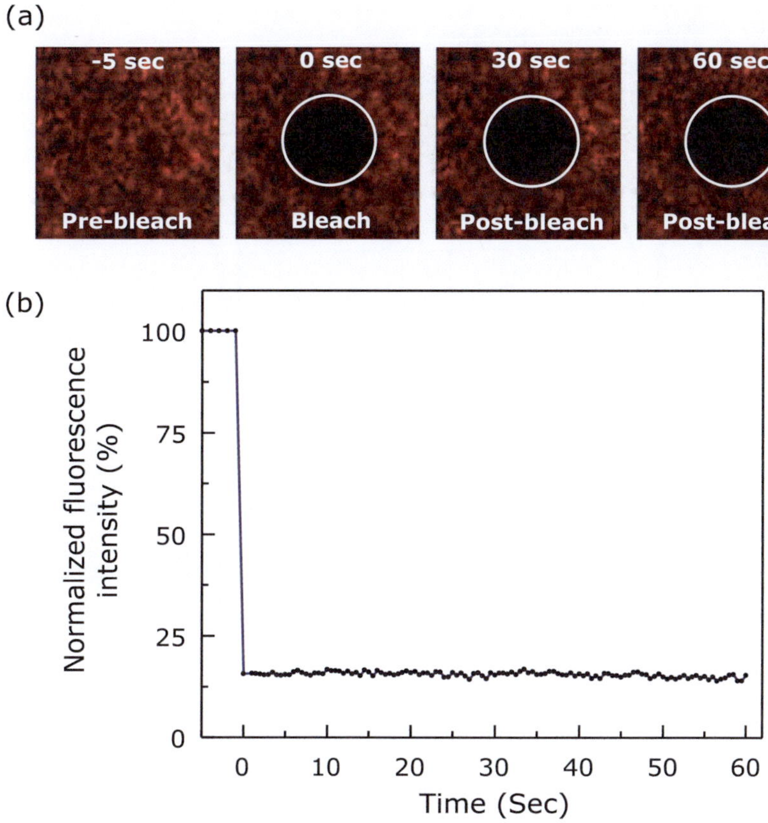

Fig. 5 Control experiments to check irreversibility of photobleaching. Reversibility of photobleaching could be monitored by bleaching a fixed sample using the identical imaging setup used for live samples. Panel (**a**) shows FRAP measurements performed on *FAST* DiI labeled plasma membrane of CHO-K1 cells that were fixed using 3.5% (v/v) formaldehyde. Panel (**b**) shows fluorescence recovery kinetics in the ROI (marked with a white circle in panel (**a**)). In absence of any reversible photobleaching, no fluorescence recovery should be observed

8. Photoreversible bleaching poses a problem during the quantitative analysis of FRAP data and could lead to an erroneous estimation of lateral diffusion [40, 41]. It is critical to establish conditions which confirm that bleaching is irreversible during the time course and condition of the experiment. To check for reversible photobleaching of the fluorophore in FRAP experiments, the bleaching conditions should be first standardized in fixed samples in which no recovery of fluorescence should be expected (*see* Fig. 5).
9. A simple test for photobleaching-induced immobile fractions due to crosslinking of fluorophores is to perform multiple FRAP measurements in the same region of interest in the same cell [42, 43] (Fig. 6). For such measurement, the diffusion coefficient should not change but the mobile fraction should be close to 100% in subsequent FRAP measurements

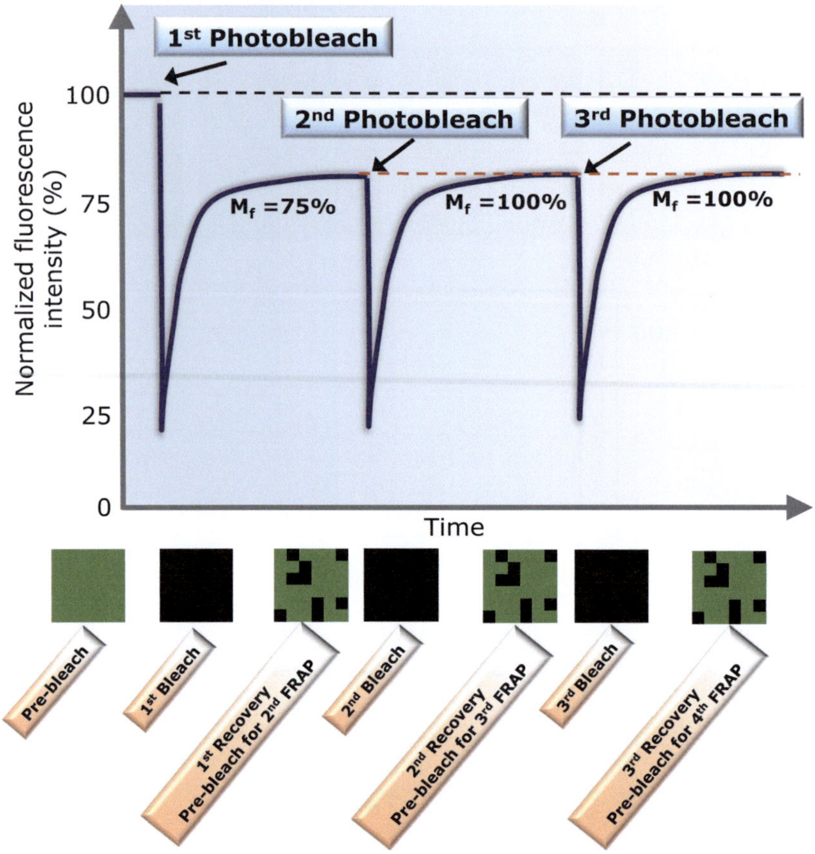

Fig. 6 Control experiments to check for photo-induced immobile fractions. A simple test for generation of photo-induced immobile fractions is to perform a second FRAP measurement in the same ROI in the same cell. In this example, the mobile fraction of the first FRAP experiment is ~75%. In the absence of photodamage-induced artifacts, complete fluorescence recovery should be observed from subsequent FRAP measurements in the same ROI. The bottom panel shows an illustration of time evolution of the bleach ROI during multiple photobleaching FRAP measurements. The black spots represent bleached fluorophores and the green spots represent unbleached fluorophores. In case of an ideal FRAP experiment, the post-recovery intensity after the second FRAP measurement should be the same as the pre-bleach intensity of the given FRAP measurement. A high recovery (~100%) during subsequent FRAP measurements indicates the actual "immobile" fraction, whereas reduced mobile fraction (comparable to first FRAP measurement) indicates potential photodamage and could lead to artifacts in data analysis

in the same ROI. As shown in Fig. 6, the mobile fraction of the first FRAP experiment is ~75%. In the absence of photodamage, 100% fluorescence recovery should be observed from second and subsequent FRAP measurements. The immobile molecules will be photobleached in the first FRAP and therefore would not contribute to the percent recovery observed in subsequent FRAP measurements. A lower value of mobile fraction (<100%) from subsequent FRAP experiments would

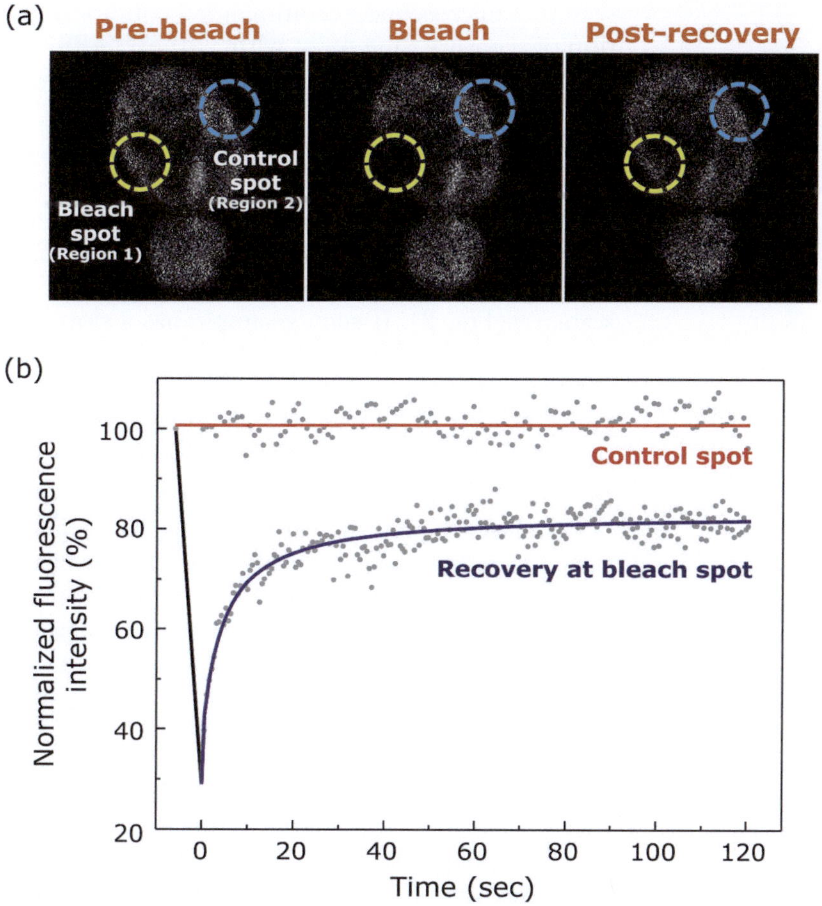

Fig. 7 A control experiment to check for photobleaching during imaging. (**a**) A representative FRAP experiment performed on *Candida* cells labeled with *FAST* DiI. The fluorescent periphery of cells representing the plasma membrane was selected for bleaching and monitoring recovery of fluorescence. Confocal images of the same cell before bleach (pre-bleach), immediately after bleach (bleach), and after recovery (post-recovery) for a representative FRAP experiment are shown in (**a**). Region 1 (bleach spot shown as yellow dashed line; radius = 1 μm) was monitored to measure fluorescence recovery after photobleaching; region 2 (control spot shown as blue dashed line) to detect any possible bleaching during scanning. Data representing the normalized mean fluorescence intensities from the regions depicted in (**a**) for the entire duration of the FRAP experiment are shown in (**b**). The relatively constant fluorescence intensity in the control spot (region 2) in (**b**) shows that the imaging conditions were optimized, and no significant photobleaching of fluorescence during time-lapse imaging took place. The solid lines are nonlinear regression fits of the data to Eq. (1). Adapted and modified with permission from [44]

indicate crosslinking-induced generation of immobile fluorophores.

10. A quick check to make sure that there is no photobleaching during imaging is to monitor the fluorescence intensity profile of a region away from the actual bleach spot (*see* Fig. 7). Fluorescence intensity of such a control spot should remain invariant with time. In a Zeiss LSM 510 meta confocal laser

scanning microscope, recommended condition for the acquisition of recovery time point with a 40 mW 488/514 nm argon laser line is 30–50% laser power with 0.1% to 2% transmission, for a specific pinhole (1 Airy unit).

11. In control experiments, a sample should be imaged using identical imaging settings without bleaching the ROI to find out the maximum number of scans possible before significant photobleaching of the sample. Let us assume that for a particular sample, photobleaching is observed after n number of scans (*see* Fig. 8). If such a sample has a slow rate of fluorescence recovery due to slow diffusion, a longer delay time (between two successive scans) should be given to capture the entire process of fluorescence recovery (till recovery reaches a plateau) within n number of scans (*see* Fig. 8).

12. In any FRAP measurement, it is crucial to capture the recovery of fluorescence intensity with highest possible temporal resolution. Currently, most of the commercially available laser scanning confocal microscopes are equipped with bidirectional scanning. Bidirectional scanning allows imaging the sample from both directions along the x-axis and significantly increases the scan rate. However, it is often necessary to adjust the phase shift correction to avoid image distortion. Additionally, line/frame averaging should be avoided while imaging as it slows down scan speed and leads to loss of time resolution between successive scans. For rapidly diffusing molecules where very high temporal resolution is required to capture the fluorescence recovery kinetics, it is advised to image the bleach ROI alone for rapid collection of data points. The only caveat of this approach is that the cell may shift in position during the course of time-lapse imaging, and this could induce further complication.

13. Discard fluorescence recovery traces that exhibit the following artifacts: (a) unstable pre-bleach fluorescence intensity, (b) recovery curves that do not display a plateau, (c) photobleaching during acquisition, and (d) fluctuations of fluorescence intensity due to cell movement (*see* Fig. 9).

14. Precise determination of the diffusion coefficient from a FRAP measurement depends strongly on how similar the dimension of the theoretical ROI (ω) is to the actual bleach spot dimension in the sample (since D has a square dependence on ω). The actual size of the bleach spot may not be same as the dimensions of the ROI for all experiments. If the bleach duration is long compared to τ_d, especially for confocal microscopes with relatively low-power laser and photostable fluorophore, the effective size of the bleach spot would depend on the duration of bleach. A longer bleach duration to obtain significant

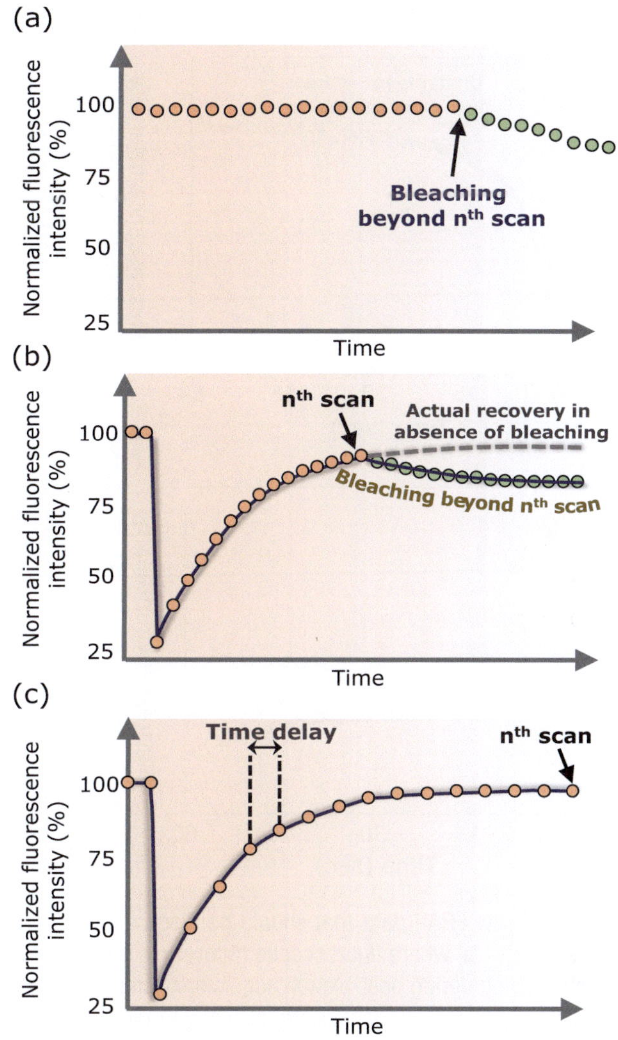

Fig. 8 Control experiment to minimize photobleaching during imaging. (**a**) A sample is imaged using identical imaging settings to be used for actual FRAP measurement, without bleaching the ROI. This would help to find out the maximum number of scans before significant photobleaching of the sample. Panel (**b**) shows that beyond *n*th scan, the post-bleach fluorescence recovery intensity is significantly affected by photobleaching. (**c**) A longer delay time (interval between two successive scans) should be given to capture the entire process of fluorescence recovery (till recovery reaches a plateau) within *n* scans

photobleaching in a sample with high diffusion tends to broaden the bleach spot (as shown in Fig. 4) leading to an underestimation of D. Fortunately, these estimates could be corrected to a significant extent by calculating the effective bleached spot size by relatively simple image analysis approaches previously described by us [45] (Fig. 10).

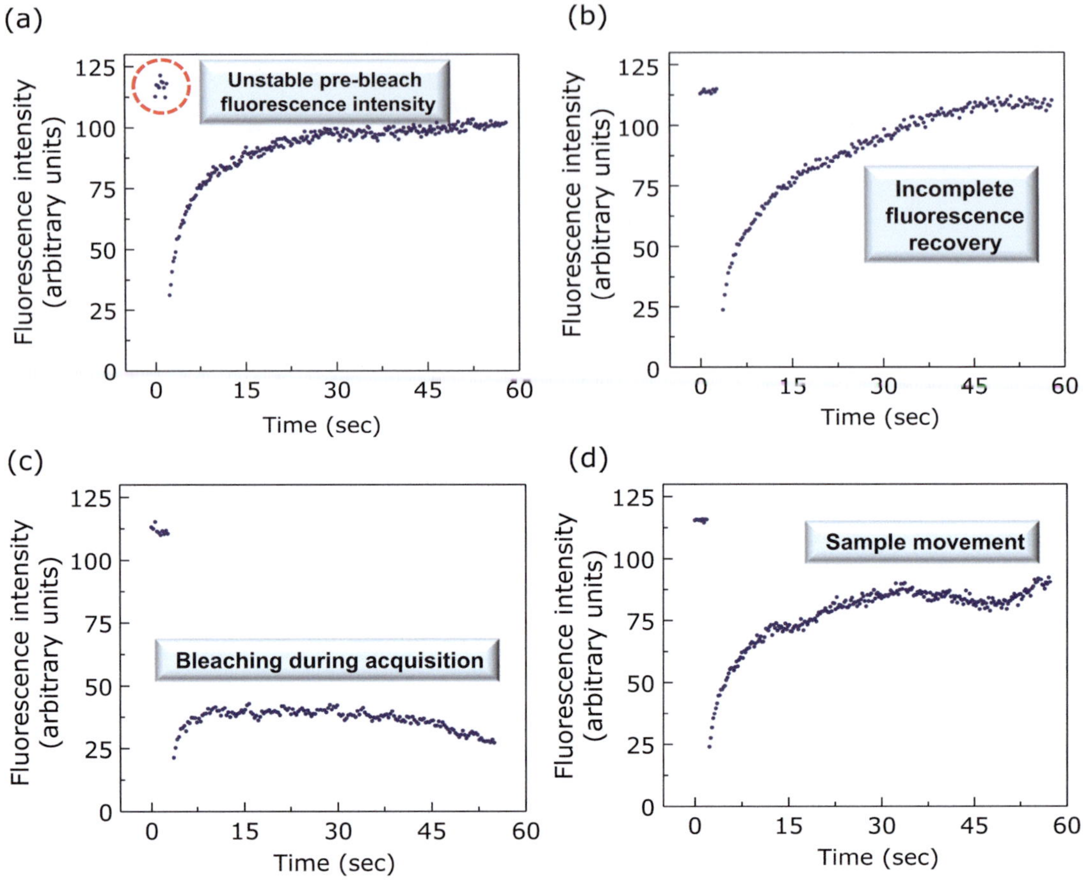

Fig. 9 Examples of raw FRAP data that should be discarded. Recovery curves (**a**) with unstable pre-bleach fluorescence intensity, (**b**) where fluorescence recovery is not complete and recovery plateau is not achieved, (**c**) with photobleaching during post-bleach acquisition, and (**d**) showing fluctuations in fluorescence intensity due to cell movement or drift of focus. Data from any of these cases should not be analyzed for calculation of diffusion parameters

15. Photobleaching using high laser illumination could lead to local heating of the sample as the laser power may reach ~1 mW/cm^2 [46, 47]. This could give rise to photo-induced crosslinking of molecules and significantly slow down diffusion. Such a phenomenon could be ruled out by performing the bleaching step using a range of laser powers. In the absence of any photo-induced crosslinking, *D* value should remain invariant over the range of laser powers used for photobleaching (*see* Fig. 11). A decrease in *D* value with higher laser power would indicate a reduction in the mobility of molecules due to the formation of higher molecular weight complex via crosslinking.
16. Equation (2) assumes unrestricted diffusion in a circular bleached ROI without any recovery from below or above the

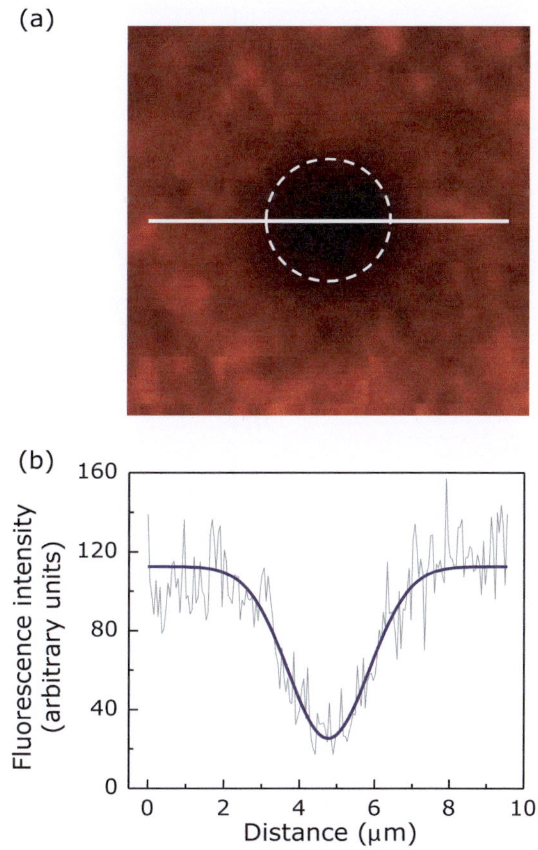

Fig. 10 Determination of actual bleach spot radius. Panel (**a**) represents the first acquired post-bleach image of *Candida* cell membrane stained with *FAST* DiI with an ROI of 1 μm radius. The fluorescence intensity profile along the straight line drawn across the bleached spot is represented in panel (**b**). The smooth line in panel (**b**) is a nonlinear regression fit of the data, and the noisy line represents a typical fluorescence intensity profile across the bleached spot. The half-width at half-maximum (HWHM) of the Gaussian amplitude function could be extracted to experimentally determine the actual bleach spot size

focal plane and is only valid for lateral diffusion in membranes [48].

17. The theoretical *D* of a protein is related to its size (mass) and its cellular environment. Deviations from this theoretical value could provide novel insight into the environment of the protein. If *D* is significantly lower than what is predicted based on the protein's mass, the protein could be incorporated in an aggregate or a large complex. In addition, the local environment of the protein could be significantly more viscous than expected. A third possibility could be that the protein could interact with a fixed scaffold such as the actin cytoskeleton [17, 49, 50]. On the other hand, if the lateral diffusion of a

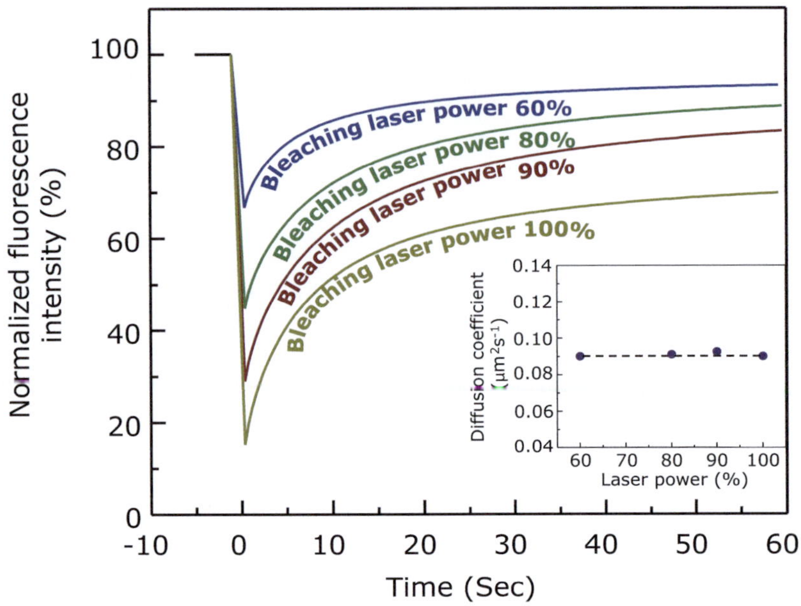

Fig. 11 A test for photo-induced crosslinking due to high bleaching laser power. Photobleaching using high laser illumination could lead to photo-induced cross-linking due to local heating and may slow down diffusion. Such a phenomenon should be ruled out by performing the bleaching step in FRAP measurements using a range of laser powers. The figure shows fitted fluorescence recovery curves to FRAP measurements performed in the same cell with a wide range of bleaching laser power. The inset shows the calculated value of the diffusion coefficient (D) for each of the fluorescence recovery curves. In absence of any photo-induced crosslinking, D value should remain invariant over the range of laser powers used for photobleaching

protein is faster than what is predicted, the protein might be undergoing directed transport via motor proteins or the local environment of the protein could be less viscous [51].

18. The mobile fraction (M_f) also reveals important information about protein dynamics. A decrease in M_f relative to control value indicates that the protein could be confined in compartments (domains) due to the formation of immobile aggregates by binding to a fixed/anchored substrate [52] and cannot contribute to fluorescence recovery. An increase in M_f would suggest that the protein is released from a fixed scaffold or exported out of a discontinuous compartment.

Acknowledgments

A.C. gratefully acknowledges support from SERB Distinguished Fellowship (Department of Science and Technology, Govt. of India). P.S. thanks the Council of Scientific and Industrial Research for the award of a Shyama Prasad Mukherjee Fellowship. A.C. is a Distinguished Visiting Professor at the Indian Institute of Technology Bombay (Mumbai), Adjunct Professor at the RMIT University (Melbourne, Australia), Tata Institute of Fundamental Research (Mumbai), and Indian Institute of Science Education and Research (Kolkata), and an Honorary Professor at the Jawaharlal Nehru Centre for Advanced Scientific Research (Bengaluru). We thank members of the Chattopadhyay laboratory for their comments and discussions.

Conflicts of Interest

There are no conflicts of interest to declare.

References

1. Edidin M (1994) Fluorescence photobleaching and recovery, FPR, in the analysis of membrane structure and dynamics. In: Damjanovich S, Edidin M, Szöllősi J, Trón L (eds) Mobility and proximity in biological membranes. CRC, Boca Raton, pp 109–135
2. Lippincott-Schwartz J, Snapp E, Kenworthy A (2001) Studying protein dynamics in living cells. Nat Rev Mol Cell Biol 2:444–456
3. Kenworthy AK (2007) Fluorescence recovery after photobleaching studies of lipid rafts. In: McIntosh TJ (ed) Lipid Rafts, Methods in molecular biology, vol 398. Humana, Totowa, pp 179–192
4. Carisey A, Stroud M, Tsang R, Ballestrem C (2011) Fluorescence recovery after photobleaching. In: Wells CM, Parsons M (eds) Cell migration: developmental methods and protocols, Methods in molecular biology, vol 769. Humana, New York, pp 387–402
5. Jafurulla M, Chattopadhyay A (2015) Novel insights in membrane biology utilizing fluorescence recovery after photobleaching. Adv Exp Med Biol 842:27–40
6. Sarkar P, Chattopadhyay A (2019) Exploring membrane organization at varying spatiotemporal resolutions utilizing fluorescence-based approaches: implications in membrane biology. Phys Chem Chem Phys 21:11554–11563
7. Lippincott-Schwartz J, Snapp EL, Phair RD (2018) The development and enhancement of FRAP as a key tool for investigating protein dynamics. Biophys J 115:1146–1155
8. Poo M, Cone RA (1974) Lateral diffusion of rhodopsin in the photoreceptor membrane. Nature 247:438–441
9. Liebman PA, Entine G (1974) Lateral diffusion of visual pigment in photoreceptor disk membranes. Science 185:457–459
10. Peters R, Peters J, Tews KH, Bähr W (1974) A microfluorimetric study of translational diffusion in erythrocyte membranes. Biochim Biophys Acta 367:282–294
11. Axelrod D, Koppel DE, Schlessinger J, Elson E, Webb WW (1976) Mobility measurement by analysis of fluorescence photobleaching recovery kinetics. Biophys J 16:1055–1069
12. Jacobson K, Derzko Z, Wu E-S, Hou Y, Poste G (1976) Measurement of the lateral mobility of cell surface components in single, living cells by fluorescence recovery after photobleaching. J Supramol Struct 5:565(417)–576(428)
13. Chalfie M, Tu Y, Euskirchen G, Ward WW, Prasher DC (1994) Green fluorescent protein as a marker for gene expression. Science 263:802–805

14. Reits EAJ, Neefjes JJ (2001) From fixed to FRAP: measuring protein mobility and activity in living cells. Nat Cell Biol 3:E145–E147
15. Lagerholm BC, Weinreb GE, Jacobson K, Thompson NL (2005) Detecting microdomains in intact cell membranes. Annu Rev Phys Chem 56:309–336
16. Valdez-Taubas J, Pelham HRB (2003) Slow diffusion of proteins in the yeast plasma membrane allows polarity to be maintained by endocytic cycling. Curr Biol 13:1636–1640
17. Ganguly S, Pucadyil TJ, Chattopadhyay A (2008) Actin cytoskeleton-dependent dynamics of the human serotonin$_{1A}$ receptor correlates with receptor signaling. Biophys J 95:451–463
18. Tsien RY (1998) The green fluorescent protein. Annu Rev Biochem 67:509–544
19. White J, Stelzer E (1999) Photobleaching GFP reveals protein dynamics inside live cells. Trends Cell Biol 9:61–65
20. Lippincott-Schwartz J, Patterson GH (2003) Development and use of fluorescent protein markers in living cells. Science 300:87–91
21. Haldar S, Chattopadhyay A (2009) Green fluorescent protein: the molecular lantern that illuminates the cellular interior. J Biosci 34:169–172
22. Sarkar P, Chattopadhyay A (2018) GFP fluorescence: a few lesser-known nuggets that make it work. J Biosci 43:421–430
23. Fucile S, Palma E, Martínez-Torres A, Miledi R, Eusebi F (2002) The single-channel properties of human acetylcholine α7 receptors are altered by fusing α7 to the green fluorescent protein. Proc Natl Acad Sci USA 99:3956–3961
24. Saffman PG, Delbrück M (1975) Brownian motion in biological membranes. Proc Natl Acad Sci USA 72:3111–3113
25. Maekawa M, Fairn GD (2014) Molecular probes to visualize the location, organization and dynamics of lipids. J Cell Sci 127:4801–4812
26. Klymchenko AS, Kreder R (2014) Fluorescent probes for lipid rafts: from model membranes to living cells. Chem Biol 21:97–113
27. Klausner RD, Wolf DE (1980) Selectivity of fluorescent lipid analogues for lipid domains. Biochemistry 19:6199–6203
28. Spink CH, Yeager MD, Feigenson GW (1990) Partitioning behavior of indocarbocyanine probes between coexisting gel and fluid phases in model membranes. Biochim Biophys Acta 1023:25–33
29. Kalipatnapu S, Chattopadhyay A (2004) A GFP fluorescence-based approach to determine detergent insolubility of the human serotonin$_{1A}$ receptor. FEBS Lett 576:455–460
30. Baumgart T, Hunt G, Farkas ER, Webb WW, Feigenson GW (2007) Fluorescence probe partitioning between L_o/L_d phases in lipid membranes. Biochim Biophys Acta 1768:2182–2194
31. Mukherjee S, Soe TT, Maxfield FR (1999) Endocytic sorting of lipid analogues differing solely in the chemistry of their hydrophobic tails. J Cell Biol 144:1271–1284
32. Pucadyil TJ, Chattopadhyay A (2006) Effect of cholesterol on lateral diffusion of fluorescent lipid probes in native hippocampal membranes. Chem Phys Lipids 143:11–21
33. Mukhopadhyay K, Prasad T, Saini P, Pucadyil TJ, Chattopadhyay A, Prasad R (2004) Membrane sphingolipid-ergosterol interactions are important determinants of multidrug resistance in *Candida albicans*. Antimicrob Agents Chemother 48:1778–1787
34. Soumpasis DM (1983) Theoretical analysis of fluorescence photobleaching recovery experiments. Biophys J 41:95–97
35. Lippincott-Schwartz J, Presley JF, Zaal KJM, Hirschberg K, Miller CD, Ellenberg J (1999) Monitoring the dynamics and mobility of membrane proteins tagged with green fluorescent protein. Methods Cell Biol 58:261–281
36. Yang F, Moss LG, Phillips GN Jr (1996) The molecular structure of green fluorescent protein. Nat Biotechnol 14:1246–1251
37. Prendergast FG (1999) Biophysics of the green fluorescent protein. Methods Cell Biol 58:1–18
38. Haldar S, Chattopadhyay A (2007) Dipolar relaxation within the protein matrix of the green fluorescent protein: a red edge excitation shift study. J Phys Chem B 111:14436–14439
39. Weiss M (2004) Challenges and artifacts in quantitative photobleaching experiments. Traffic 5:662–671
40. Dickson RM, Cubitt AB, Tsien RY, Moerner WE (1997) On/off blinking and switching behaviour of single molecules of green fluorescent protein. Nature 388:355–358
41. McAnaney TB, Zeng W, Doe CFE, Bhanji N, Wakelin S, Pearson DS, Abbyad P, Shi X, Boxer SG, Bagshaw CR (2005) Protonation, photobleaching, and photoactivation of yellow fluorescent protein (YFP 10C): a unifying mechanism. Biochemistry 44:5510–5524
42. Lippincott-Schwartz J, Altan-Bonnet N, Patterson GH (2003) Photobleaching and

photoactivation: following protein dynamics in living cells. Nat Cell Biol 5:S7–S14

43. Snapp EL, Altan N, Lippincott-Schwartz J (2003) Measuring protein mobility by photobleaching GFP chimeras in living cells. Curr Prot Cell Biol 21(1):1–21.1.24
44. Khandelwal NK, Sarkar P, Gaur NA, Chattopadhyay A, Prasad R (2018) Phosphatidylserine decarboxylase governs plasma membrane fluidity and impacts drug susceptibilities of *Candida albicans* cells. Biochim Biophys Acta 1860:2308–2319
45. Pucadyil TJ, Chattopadhyay A (2006) Confocal fluorescence recovery after photobleaching of green fluorescent protein in solution. J Fluoresc 16:87–94
46. Axelrod D (1977) Cell surface heating during fluorescence photobleaching recovery experiments. Biophys J 18:129–131
47. De Los SC, Chang C-W, Mycek M-A, Cardullo RA (2015) FRAP, FLIM, and FRET: detection and analysis of cellular dynamics on a molecular scale using fluorescence microscopy. Mol Reprod Dev 82:587–604
48. Koppel DE, Sheetz MP, Schindler M (1980) Lateral diffusion in biological membranes. A normal-mode analysis of diffusion on a spherical surface. Biophys J 30:187–192
49. Pucadyil TJ, Chattopadhyay A (2007) Cholesterol depletion induces dynamic confinement of the G-protein coupled serotonin$_{1A}$ receptor in the plasma membrane of living cells. Biochim Biophys Acta 1768:655–668
50. Ganguly S, Chattopadhyay A (2010) Cholesterol depletion mimics the effect of cytoskeletal destabilization on membrane dynamics of the serotonin$_{1A}$ receptor: a zFCS study. Biophys J 99:1397–1407
51. Kreutzberger AJB, Ji M, Aaron J, Mihaljević L, Urban S (2019) Rhomboid distorts lipids to break the viscosity-imposed speed limit of membrane diffusion. Science 363:eaao0076
52. Feder TJ, Brust-Mascher I, Slattery JP, Baird B, Webb WW (1996) Constrained diffusion or immobile fraction on cell surfaces: a new interpretation. Biophys J 70:2767–2773

Chapter 9

Cyclodextrins for Probing Plasma Membrane Lipids

Amid Vahedi and Amir M. Farnoud

Abstract

Cyclodextrins are cyclic molecules with a hydrophilic external structure and a hydrophobic internal cavity. Due to their chemical structure, cyclodextrins can form complexes with hydrophobic or amphiphilic molecules, including lipids. This property has resulted in the widespread use of cyclodextrins for characterization and manipulation of lipids in biological membranes. In this chapter, the use of cyclodextrins for characterization of lipids in the cell plasma membrane is discussed. First, the chemical structure of different cyclodextrins is explained. Then, the current knowledge on the mechanisms of cyclodextrin–lipid interaction and the role of cyclodextrin and lipid chemical structure in the outcome of such interactions is summarized. The use of cyclodextrins for manipulation of plasma membrane lipids and its application to study membrane lipid rafts is also reviewed. Finally, future directions in the field and current opportunities to utilize cyclodextrins to provide a better understanding of the composition and organization of lipids in the plasma membrane are discussed.

Keywords Cyclodextrin, Cholesterol, Phospholipids, Plasma membrane, Lipid rafts

Nomenclature

CD	Cyclodextrin
CEγCD	Carboxyethylated-γCD
DMβCD	Dimethyl-β-cyclodextrin
DMPC	Dimyristoyl phosphocholine
DPPC	Dipalmitoyl phosphatidylcholine
DPPE	Dipalmitoyl phosphoethanolamine
DSPC	Distearoyl phosphocholine
GUV	Giant unilamellar vesicle
HPαCD	Hydroxypropylated-αCD
HPβCD	Hydroxypropyl-βCD
LUV	Large unilamellar vesicle
MβCD	Methyl-β-cyclodextrin
PC	Phosphatidylcholine
PE	Phosphatidylethanolamine
PI	Phosphatidylinositol

Rajendra Prasad and Ashutosh Singh (eds.), *Analysis of Membrane Lipids*, Springer Protocols Handbooks,
https://doi.org/10.1007/978-1-0716-0631-5_9,

PS	Phosphatidylserine
POPC	1-Palmitoyl-2-oleoyl-glycero-3-phosphocholine
SM	Sphingomyelin
SUV	Small unilamellar vesicles
TMβCD	Trimethyl-β-cyclodextrin

1 Introduction

Cyclodextrins (CDs) are a series of cyclic oligosaccharide molecules that are composed of α-(1-4) linked D-glycopyranose units [1–3]. These cyclic compounds form a conical cylinder, best characterized as a wreath-shaped truncated cone [1]. The resulting structure has a hydrophilic external face and a hydrophobic internal cavity. The internal hydrophobic environment is the result of lining of cavity's wall with two rings of C–H groups and a glucosidic oxygen bridge ring in the middle [4]. This unique property of the CD molecules has made them a suitable choice as a carrier for hydrophobic molecules, as CDs are water soluble, but are capable of encapsulating hydrophobic molecules inside their cavity [5, 6].

The three main types of CDs, αCD, βCD, and γCD consist of six, seven, or eight D-glycopyranose units, respectively (Fig. 1). These are crystalline, homogeneous, nonhygroscopic substances naturally made by enzymatic degradation of starch [2]. As the number of D-glycopyranose units increases, the size of the hydrophobic cavity increases as well (Table 1).

This means that each type of CD can serve as a "host" for different hydrophobic molecules with its selectivity primarily based on the size of the molecule [5].

Although the chemical structures of the αCD, βCD, and γCD are similar, their water solubility differs. The solubility of βCD is approximately nine times lower than the other two. This is due to a secondary belt formation by the hydrogen bonds between the C–2–OH and C–3–OH groups of two adjacent glycopyranose units [1]. The H-bond belt in αCD is incomplete because one of the glycopyranose units is in a distorted position, and therefore the H-bond belt is weaker. γCD has a non-coplanar, more flexible structure and has the highest solubility of the three. βCD cannot be dissolved in most organic solvents but is soluble in organic solvent/water mixtures [7, 8].

Chemical modifications of the primary and secondary hydroxyl groups are commonly performed on main CD types to change certain characteristics of the molecules. These modifications include, but are not limited to, amination, methylation, esterification, and etherification [9]. Modified CD molecules are often called CD derivatives as the synthesis process starts with a main type of CD and continues to modify the hydroxyl groups. A well-known

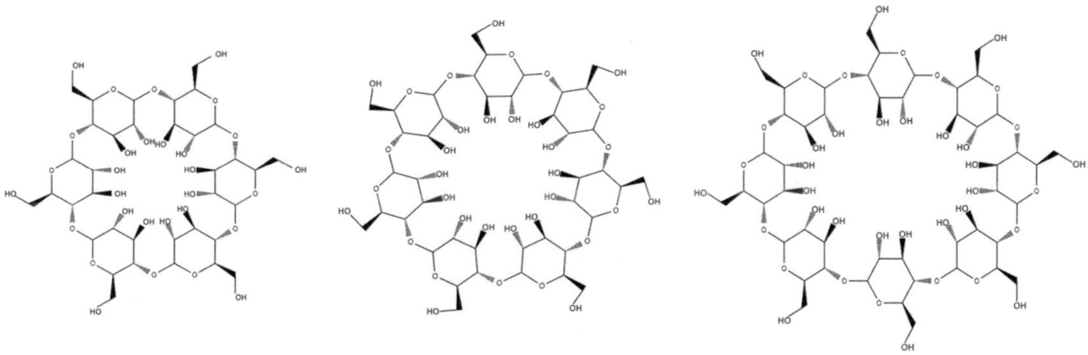

Fig. 1 Chemical structure of the three main types of CDs, α (left), β (middle), and γ (right)

Table 1
Approximate dimensions of the three main cyclodextrins (adapted from [1])

	αCD	βCD	γCD
Cavity diameter (Å)	4.7–5.3	6.0–6.5	7.5–8.3
Height of torus (Å)	7.9 ± 0.1	7.9 ± 0.1	7.9 ± 0.1
Diameter of outer periphery (Å)	14.6 ± 0.4	15.4 ± 0.4	17.5 ± 0.4

example of CD modifications is enhancing the hydrophilicity of βCD by methylation or 2-hydroxypropylation to increase its solubility in water [10].

CDs have long been used in solubilizing hydrophobic molecules, due to their ability to provide a highly hydrophobic cavity in a hydrophilic environment. This feature has been applied in studying and modulating plasma membrane lipids. In this chapter, first the interactions of CDs and their derivatives with different lipids and the parameters affecting these interactions will be discussed. Then, the applications of CDs to extract, exchange, and, generally, manipulate lipid contents in vesicular and biological membranes will be explained.

2 Interactions of Cyclodextrins with Lipids and Membranes

The interaction of lipids and CDs is known to be of the guest-host type [2, 11]. This interaction type is basically the formation of an inclusion complex, where the lipids (guest) are incorporated in the cavity of the CD (host). As mentioned earlier, the inclusion complex is highly size dependent, but other factors such as hydrophobicity play a major role as well. Almost all of the lipid classes (fatty acids, triglycerides, phospholipids, etc.) can show an interaction of this type with a CD molecule, given that the choice of CD

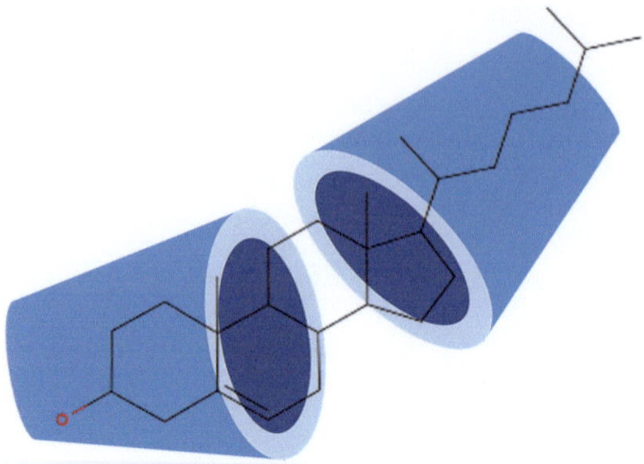

Fig. 2 Schematic illustration of the cholesterol and βCD complex. A 1:2 cholesterol:βCD molar ratio is reported based on experimental and simulation studies

molecule, i.e., the size of the cavity, is suitable for the lipid class [11]. It is noteworthy that the molar ratios of interactions are not the same for all lipid classes. Various spectroscopic methods such as nuclear magnetic resonance (NMR), UV–vis spectrophotometry, and fluorescence spectroscopy have been reported for detecting the formation of an inclusion complex [2, 5]. The mechanism behind the formation of an inclusion complex and its dependence on the CD and lipid structure has been a topic of intense research. In this section, the insights gained from studies on CD-cholesterol and CD-phospholipid interactions have been summarized.

2.1 Interaction of CDs with Cholesterol

The inclusion of cholesterol in βCD and its derivatives is one of the most studied lipid–CD inclusion complexes. This is mainly due to the important role of cholesterol in living organisms, as part of the plasma membrane and intracellular organelles, as well as the availability of βCD and its derivatives.

A good understanding of any inclusion complex is only possible if its stoichiometry is known. Many studies have investigated possible molar ratios for the inclusion complex of βCD and cholesterol. Inclusion of cholesterol inside the cavity of βCD and one of its derivatives dimethyl-β-cyclodextrin (DMβCD) was investigated using NMR and oxidation reactions. The results indicated a 1:2 cholesterol:βCD molar ratio, as two cholesterol rings are encapsulated in the cavity of one βCD molecule and one ring and the side chains are encapsulated in the other [12]. This is shown schematically in Fig. 2. Molecular dynamics (MD) simulations also suggest two local minima in free energy profiles for the inclusion of cholesterol in βCD and two of its derivatives. These two local minima can be attributed to the probable binding modes for the inclusion complex [13]. Further investigation of inclusion complex formations of charged and uncharged βCD derivatives also suggest a 1:2 cholesterol:βCD derivative molar ratio for uncharged derivatives but are inconclusive for the charged βCD derivatives [14].

MD simulations and implementing free energy measurements in different scenarios provide a facile method for stepwise investigation of the mechanisms of inclusion complex formation between cholesterol and βCD. A recent study proposes a five-step mechanism based on this method: βCD dimer formation, binding of the dimers on the membrane, inclusion of the cholesterol in the cavity of the βCD molecule, tilting of the inclusion complex, and desorption of the complex from the membrane [15]. It is important to mention that the inclusion of cholesterol into the cavity of a CD molecule has a lower activation energy compared to other methods of cholesterol efflux from the cell membrane [16]. In cholesterol efflux without a CD molecule, cholesterol is first desorbed completely into the aqueous phase and then absorbed by HDL particles or phospholipid liposomes. With a CD molecule present at the interface, the desorption into aqueous phase step is removed, as cholesterol transfers directly from the membrane into the cavity of a CD molecule. The activation energy is lower in the latter case, and therefore, it will be the preferred mechanism.

2.2 Interaction of CDs with Phospholipids

Investigating the interactions of CDs with phospholipids is not as straightforward as it is for cholesterol. This is because phospholipids are a vast variety of molecules with different characteristics. Any investigation into such interactions should consider intrinsic characteristics of phospholipids, such as the different headgroups, acyl chain lengths, and saturation states. Such studies are further complicated by the fact that any study on CD-phospholipid interactions would need to consider the role of the membrane models (e.g., micelles, monomers, bilayers, LUVs, or GUVs) [17].

In one of the first studies in the field, interactions of different CDs and some of their derivatives with cell membrane phospholipids was investigated using pure phospholipids (DPPE and DPPC) dissolved in organic solvent, liposome suspensions (DPPC) in aqueous medium, and erythrocytes. The release of radiolabeled KCl from the lumen of liposomes interacting with CDs was measured as an indicator for membrane permeabilization. The results indicated that methylated βCDs increased the permeability of liposome models, as evidenced by an increase in the release of radiolabeled KCl. Hemolysis (disruption of erythrocytes and release of hemoglobin) only occurred at high concentrations of βCD and MβCD likely due to the formation of CD-lipid inclusion complexes [18].

Acyl chain length plays a major role in the interaction of CDs with phospholipids. A simple approach for studying the effects of acyl chain length is to use alkane molecules. The association of normal alkanes (C_5–C_9) with different CDs was investigated using the volatilization rate of alkane molecules. The results indicate that

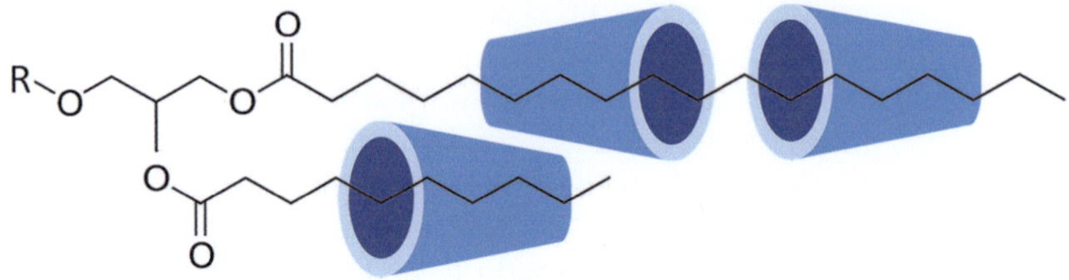

Fig. 3 Schematic illustration of the inclusion complex formed upon the interaction of phospholipids with αCDs. Note that the exact stoichiometry depends on the structure of the lipid and CD

although a 1:1 alkanes:CDs molar ratio can be obtained for βCD and γCD for all alkyl chain lengths, a 1:1 and 1:2 alkanes:αCD molar ratio was obtained for short and long alkyl chains, respectively [19]. Although this study notes that, considering the cavity size, inclusion of two alkyl chains in one αCD molecule is not possible, other researchers have suggested inclusion of both acyl chains of a pyrene-labeled PC (acyl chain lengths between 6 and 14 C) in two to three γCD molecules [20].

The presence of two acyl chains in phospholipids means that more CD molecules are needed to serve as a "host" for these lipids. A calorimetric measurement of the interactions of POPC and methyl-β-cyclodextrin (MβCD) suggests a 1:4 POPC:MβCD molar ratio. Since the oleoyl carbon chain in POPC includes 18 carbons, this implies that with increased length of the acyl chain, two MβCD molecules are needed to form inclusion complexes with each chain [21]. A similar phenomenon is observed in studies on the interaction of CDs with other phospholipids, where shorter acyl chain length requires lower number of CD molecules to facilitate the interaction [17, 20]. This process is shown schematically in Fig. 3. However, even for lipids with the same acyl chain length, the outcome of CD-lipid interactions highly depends on the structure of the CD. A study on the interactions of CDs with PC liposomes, monitored by the efflux of a self-quenching fluorescent dye, carboxyfluorescein (CF), encapsulated in the liposomes, clearly shows the importance of CD structure. In this study, liposomes composed of DMPC (14:0), DPPC (16:0), and DSPC (18:0) were exposed to αCD, dimethyl-β-cyclodextrin (DMβCD), and trimethyl-β-cyclodextrin (TMβCD) and the leakage of CF was examined. Although the CF release profiles were dependent on the concentration of CDs, as the alkyl chain length of the lipids increased, for αCD, DMβCD, and TMβCD, the CF release decreased, increased, and was relatively constant, respectively [22]. Based on differential scanning calorimetry (DSC) studies, the investigators proposed that DMβCD is adsorbed on the liposome surface and penetrates into the matrix of the liposome, while αCD is adsorbed on the liposome, but cannot penetrate the lipid bilayer and will extract DPPC molecules from the surface [22]. The differences in the mechanisms of CD–lipid interactions could explain the differences in the leakage response once lipids of the same chain length are exposed to different CDs. It should be noted that not only the acyl chain length but also the saturation state of the acyl chain should be considered in understanding the interactions. CDs interact better with saturated phospholipids as the double bonds in unsaturated acyl chain impose an obstacle for "sliding" of the αCD molecule up the acyl chain [23, 24].

Lipid headgroup is another important factor in the interactions of CDs with phospholipids. NMR was used to investigate the interaction of αCD with small unilamellar vesicles (SUVs) of two relatively similar, naturally occurring phospholipids with different headgroups (egg PC and PI). The results indicate quite different interactions for the two lipids: no interactions with PC and strong interactions with PI. The difference in the interactions is attributed to the specific recognition of headgroups by CDs followed by nonspecific inclusion complex formation with acyl chains [25]. When different phospholipid headgroups (PC, PI, PS, and PE), with the same acyl chain structure, were compared for their interaction with αCD using NMR and MD simulations, it was found that the inositol in PI interacts strongly with αCD and is either completely included in its cavity or is associated with it through hydrogen bonding [24]. This type of interaction with αCD was not reported for other headgroups. Interaction of acyl chain of PI with αCD is also different, as the results indicate inclusion of the terminal methyl group of the acyl chain in the cavity of αCD; an interaction that was not recorded for other phospholipids [23, 24]. However, in phospholipids other than PI, the headgroup still plays a role in dictating the outcome of lipid–CD interactions, despite the fact that interactions primarily occur through inclusion complex formation with the acyl chain. For example, a study with various methylated CDs has shown that PC and SM interact more effectively with CDs compared to PS and PE [17].

In summary, while the details of the interactions depend on the structure of the phospholipid and the CD, several general statements regarding the role of the phospholipid structure in determining the outcome of lipid–CD interactions can be made. Phospholipids interact with CDs through formation of acyl chain inclusion complexes with the cavity of CDs. The exception to this statement is PI, which shows headgroup interaction with the cavity of CDs as well as acyl chain inclusion. CDs are capable of forming inclusion complexes with both acyl chains of the phospholipids, and as the acyl chain length increases, the number of CD molecules increases as well. The lipid headgroup does play a role in this process, with certain headgroups facilitating the interactions better than others. Finally, the double bond of an unsaturated acyl chain can cause an obstacle in the formation of an inclusion complex.

3 Cyclodextrins for Manipulation of Lipids in the Plasma Membrane

Modulation of cellular lipids by CDs has been widely used to uncover the role of specific lipids in regulating cellular processes. It should be noted that the use of CDs to modulate lipid composition is not limited to mechanistic studies of plasma membrane biology. CDs have been used to modulate lipid levels as therapeutic agents in cardiovascular or neurological diseases (*see*, e.g., [26]). However, this section is focused on studies that have utilized CDs to probe biophysical or biochemical processes in cellular membranes. Emphasis has been given to cholesterol and phospholipids, which are the abundant lipids in the plasma membrane and are primary targets for modulation with CDs. Since CDs can be used to both extract and replace or incorporate lipids in the cell membrane, these processes have been discussed separately.

3.1 CDs for Alteration of Cholesterol in Plasma Membranes

3.1.1 CDs for Cholesterol Extraction

Extraction of cholesterol from the cell plasma membrane has been widely used to study ordered lipid domains, commonly known as lipid rafts. Lipid rafts are defined as "small (10–200 nm), heterogeneous, highly dynamic, sterol and sphingolipid-enriched domains that compartmentalize cellular processes" [27]. Given the importance of sterols in the formation of rafts, cholesterol extraction has become an important method of studying the role of lipid rafts in cellular processes.

βCD and its derivatives interact well with cholesterol, and exposing cells in vitro to βCDs results in cholesterol removal from the cell. Numerous researchers have studied the parameters affecting cholesterol depletion from membranes of different cell types. A multitude of factors, including the type and concentration of the CD, incubation time, temperature, and cell type can all affect the amount of cholesterol that can be removed from cellular membranes [5]. The most effective βCD derivative in cholesterol depletion is MβCD. Higher concentrations of MβCD and increased incubation periods are generally more effective in cholesterol depletion [5]. Interestingly, incubating inclusion complexes of cholesterol and βCDs with cells can result in enrichment of cholesterol in the cellular plasma membrane. In fact, it has been reported in multiple studies that the molar ratio of cholesterol:βCD determines whether cholesterol depletion or enrichment will take place [4, 5, 9]. The cholesterol depletion process is known to be reversible, meaning that an enrichment step following the depletion process can replenish the cholesterol contents back to normal levels [5].

The ability of βCD and its derivatives to alter the cholesterol content of the plasma membrane has been used in many studies to investigate the effects of cholesterol on cellular membrane characteristics. In a study, cellular cholesterol efflux from three different cell types was examined using different cholesterol acceptors (βCD, MβCD, 2-hydroxypropyl-βCD (HPβCD), and high density lipoprotein (HDL_3)). The results indicated that, no matter the choice of acceptor, over a period of 8 h, 50–90% of the cellular cholesterol could be depleted. However, βCD and its derivatives depleted most of the cholesterol in the first 2 h, while same level of depletion using HDL_3 took 8 h. The efflux rate constants using βCDs were 3.5- to 70-folds higher than the HDL_3. This rate was relatively constant for βCD-mediated depletion for different cell types, but when HDL_3 was used as cholesterol acceptor, the rate varied greatly [28]. It should be noted that βCD can also extract phospholipids, but this extraction is reported to be 3% of cellular phospholipids while cholesterol removal was 75% of cellular cholesterol [29].

The mechanism of cholesterol depletion from the plasma membrane using CDs has been investigated in detail using several methodologies. Kinetic analysis and MD simulations of cholesterol efflux are two of the most frequently used techniques for such studies. Most of the studies implementing kinetic analysis technique indicate that the cellular-free cholesterol exists in two pools and that the time constants for cholesterol removal from one pool are different from the other. Although investigations on this issue report differences in the exact time constants for the process, they all agree that depletion from one of the pools occurs rapidly (a few seconds), while depletion from the other one occurs slowly (15–25 min). There is, however, at least one study reporting the existence of only one cholesterol pool [30]. The discrepancies in the reports might be due to the choice of cells and their inherent differences [31].

Much argument has been made about the nature of these two pools. Both kinetic analysis of cholesterol depletion from LUVs and fluorescent microscopy analysis of supported lipid bilayers show the presence of a slow and a fast removal rate [29, 32], suggesting that the cholesterol pools should be in the bilayer membrane of the cells rather than the cytoplasmic organelles. Some investigators have attributed the two pools to the raft domains and non-raft fractions of the membrane [33]. However, detailed studies on intracellular cholesterol transport show that a large amount of cellular cholesterol is in the endocytic recycling compartment (ERC), suggesting the slow pool might be attributable to the ERC [34]. While there is debate on the nature of the slow pool, all studies agree that the fast pool corresponds to the abundant cholesterol in the outer leaflet of the plasma membrane [5].

The abundance of cholesterol within the lipid raft fractions and its convenient manipulation using βCDs has driven a lot of studies in the field, so much so that CDs are known as a lipid raft disrupting agent and sensitivity to CDs had been used as a method of lipid raft verification [35]. However, a few cautionary points should be considered in such studies: (1) a cholesterol-dependent cellular function might not necessarily be a raft-dependent function; therefore, any connection between lipid rafts and cellular functions needs to be confirmed by methods in addition to βCD-mediated cholesterol depletion; (2) βCD could extract cholesterol from both the plasma membrane and intracellular membranes with potential downstream effects not necessarily caused by raft disruption; and (3) βCD could extract cholesterol from the non-raft portions of the plasma membrane, which might not provide information on the role of lipid rafts. The latter issue, in particular, has been a matter of controversy among researchers [36]. Some researchers have reported that in the case of depletion with CDs, the cholesterol in the non-raft fractions is more likely to be removed [37]. It has also been shown that cholesterol depletion with MβCD decreases the viscous drag on the proteins associated with the lipid rafts to the level of non-raft associated proteins but does not affect the drag on the latter proteins, suggesting that CDs primarily affect raft fractions [36]. This idea has been disputed by at least one other study, in which treatment with CDs is reported to affect cholesterol contents of both raft and non-raft fractions, so that preferential depletion might not be possible [38].

Perhaps the most convincing evidence for βCD-mediated cholesterol depletion from non-raft phases has come from MD simulations in which monitoring raft and non-raft phases is easier than it is in experiments. A recent study using MD simulations investigated cholesterol extraction from lipid ordered (L_O) and lipid disordered (L_D) phases of planar and vesicular bilayer membranes made of PC/SM/cholesterol. The phases were mimicked by a tenfold difference in cholesterol concentration, with the L_O phase being the cholesterol-enriched phase. The results indicate that although βCD molecules are adsorbed roughly at the same amounts on both phases for both type of the membranes, cholesterol is preferably extracted from the L_O phase [39].

3.1.2 CD-Mediated Sterol Exchange

βCDs have been utilized to exchange cholesterol with other sterols in cellular membranes. This process facilitates studies on the role of cholesterol in cells and whether these roles can be filled with other sterols. Effects of other sterols replacing cholesterol in J774 macrophages was investigated utilizing βCD-mediated exchange of cholesterol with desmosterol in a two-step process. The cells were first depleted of their cholesterol using HPβCD supplemented with cholesterol biosynthesis inhibiting compounds in the media, and then using MβCD complexation, desmosterol was introduced to the cellular membrane. As J774 cells are known to be unable to convert desmosterol to cholesterol, the indefinite growth of cells was considered to show that desmosterol can continue cholesterol functions in cells [40]. In another study, effects of different sterol structures on the endocytosis capabilities of MDA-MB-231 breast cancer cells were investigated. Twelve different sterols were separately used to replace cellular cholesterol by incubating cells with MβCD complexed with sterols. Using thin-layer chromatography and mass spectroscopy, it was confirmed that exogenous sterols were exchanged with cellular cholesterol. The results indicated that the ability of sterols to support lipid raft formation is necessary but not sufficient for endocytosis [41]. A recent investigation on the effect of different sterols on the autophosphorylation of insulin receptor and insulin-like growth factor 1 receptor in HEK 293 T cells showed that the ability to form ordered domains is sufficient for activating the mentioned receptors in a cholesterol-depleted cell [42]. Together these studies suggest that the sterols with the ability to support domain formation might be able to carry out the role of cholesterol in the cell.

3.2 CDs for Alteration of Phospholipids in Plasma Membranes

3.2.1 CDs for Phospholipid Extraction

Studying phospholipid alterations in the cell plasma membrane is complicated. The variety of cell types, CDs and their derivatives, as well as differences in phospholipids makes it difficult to provide a comprehensive understanding of phospholipid alterations. In addition, the choice of CD to modulate membrane phospholipids has been a matter of debate. Selective interaction of cellular plasma membrane lipids with CDs is the aim of many investigations. Some studies report the extraction of sphingomyelin and phospholipids when using the CD with a small cavity, αCD. However, others have reported extraction of phospholipids, as well as cholesterol, when cells are interacting with βCDs [5]. In an early investigation, using erythrocytes, it was reported that αCD and γCD are capable of changing the shape of the cells while βCD causes hemolysis before the shape change is observed. It was suggested that αCD is the most potent in solubilizing phospholipids while βCD is best in solubilizing cholesterol and proteins. It is also reported that the solubilization happens without the insertion of the CDs into the membrane [43]. In a more recent study, the effects of αCD, MαCD, and hydroxypropylated-αCD (HPαCD) on rabbit erythrocytes was investigated. The results indicate that the morphological changes and hemolytic activity of these three CDs are due to extraction of sphingomyelin and phospholipids from the plasma membrane rather than cholesterol, with MαCD showing the largest effect [44].

3.2.2 CD-Mediated Phospholipid Exchange

The cell plasma membrane is an asymmetric bilayer, and the phospholipid composition of the outer leaflet is different from the inner leaflet. While lipid vesicles are commonly used to gain mechanistic insight into lipid–lipid and lipid–protein interactions in the cell plasma membrane, the majority of commonly used models are symmetric. Naturally, the use of asymmetric vesicles would increase the biological relevance of such studies. The need to mimic the phospholipid asymmetry of the plasma membrane has been an important driver for the use of CDs to exchange phospholipids between different vesicles.

The process of CD-mediated phospholipid exchange essentially consists of the formation of CD-lipid complexes, using lipids from a donor vesicle, followed by incorporation of the lipids into an acceptor vesicle. This process has been performed using β and γCDs. An early study using carboxyethylated-γCD (CEγCD) demonstrated the feasibility of using CD for the exchange of phospholipids from one vesicle to another. It was shown that while PC, PE, and PS are exchanged with acceptor vesicles to almost the same extent, SM and PI are exchanged, respectively, three and two times more than PC, and this is attributed to differences in the hydrophobicity of SM and polarity of the head group of PI [45]. A more recent study compared the ability of various CDs to interact with phospholipids. HPαCD, HPβCD, and MβCD were reported to interact with phospholipids better than the other CDs. While phospholipid exchange capabilities were far stronger in MβCD, αCD was considered to be a better mediator of phospholipid exchange, as it does not interact with cholesterol, allowing for modulation of phospholipids without affecting cholesterol content [46]. These findings led to a number of follow-up studies in which αCDs, specifically HPαCD or MαCD, were used to prepare asymmetric phospholipids [47] to examine the role of membrane asymmetry in membrane properties such as protein localization and domain formation [47–51].

3.2.3 CDs for Phospholipid Incorporation in Cells

Incorporation or exchange of different phospholipids in cell membranes has helped researchers understand cellular membrane dynamics, roles of certain lipids in the membrane, and lipid homeostasis in general. CD-mediated exchange of phospholipids between vesicles and biological membranes will be discussed in this section.

Phospholipids were first introduced into the plasma membrane of cells using CEγCD. Pyrene-labeled PC or PS was transferred to monolayers of human fibroblasts with CEγCD as a mediator. Fluorescence microscopy of the cells incubated with pyrene-labeled lipids in the absence of CEγCD showed no fluorescence signal, suggesting that when the mediator is not present, phospholipids are not incorporated into the plasma membrane. In the presence of CEγCD, as the hydrophobicity of the pyrene-labeled PC increased (increasing the length of the acyl chain from 6 to 14 C), the rate of lipid transfer to the cells decreased to the point of no fluorescent signal. The investigation of distributed pyrene-labeled PS inside the cells showed that highly hydrophobic PS (more similar to the natural PS in the cells) remained in the plasma membrane and was not transferred to the Golgi apparatus or mitochondria [45]. In another investigation, the metabolism of individual phospholipid species in cells was studied by the incorporation of radiolabeled phospholipids in the plasma membrane of cells. Investigators synthesized 40 different species of deuterium-labeled PE and PS and introduced them into BHK21 and HeLa cells using MβCD. The donor vesicles were made of POPC/cholesterol and the exchanging PE or PS species. It was argued that the PE and PS species were most probably incorporated in the outer leaflet of the plasma membrane and that the elevation of PC levels due to incorporation of POPC was regulated after 24 h. While MβCD was capable of extracting cholesterol while delivering phospholipids, the inclusion of cholesterol into the donor vesicles allowed for maintaining a constant cholesterol concentration in the cell during the exchange process [52].

Parameters that affect phospholipid transfer from donor vesicles to cells using MβCD as a mediator have also been investigated. In one study, three different lipid headgroups (PC, PE, and PS) with different acyl chain lengths and different saturation states were transferred into four different cell lines. The results indicated that for most phospholipids, the transfer to cells is significantly dependent on the concentration of donor vesicles, incubation time, and MβCD concentration. Also, consistent headgroup dependency was not observed. However, a significant dependency on the acyl chain length and saturation state was reported. It should be mentioned that during the incorporation process, about 30% of total cellular SM was extracted; however, SM levels were normalized after 24 h [53]. In a recent investigation, the sphingolipids and phospholipids of the plasma membrane of A549 cells (as well as three other cell lines, MDA-MB-231, COS 7, and BxPC-3 cells) were replaced with exogenous lipids using MαCD as the exchange mediator. The exchange process with MαCD did not affect the cholesterol levels in the plasma membrane. It was shown that even unnatural lipids ($C_{17:0}$ SM) can be incorporated in the plasma membranes of the cells. Using radiolabeled and fluorescent lipids as well as mass spectroscopy, it was shown that the lipid exchange process is restricted to the outer leaflet lipids of the plasma membrane and that the inner leaflet lipids (PS and PI) were not exchanged [54].

4 Interactions of CDs with Membrane Proteins

Although the focus of this chapter is on membrane lipids, one cannot study the effects of a molecule on cellular membranes without considering the proteins being affected. CDs are reported to extract proteins from the membrane of different cells. A study on the interaction of CDs with erythrocytes reports that βCD and αCD are the most and least potent CDs in extracting proteins from the membrane. The results indicated that 77% of the polypeptides of bands V and VI were extracted by βCD while only about 5% of the polypeptides of bands III and IV were extracted [43]. As CDs did not increase the solubilization of pure polypeptides, the study suggested that CDs interact with lipids and not with proteins. The protein extraction in the vicinity of CDs is proposed to be an extrusion process; i.e., the proteins are secreted from the membrane that is eroded by CDs [43]. The results of another study on the effects of αCDs on the membrane of rabbit erythrocytes reported the release of certain membrane-related proteins. The most potent αCD derivative was MαCD and the least potent was HPαCD. It was also suggested that the morphological changes of the erythrocytes after treatment with αCDs might be due to the changes caused in the lipid-protein interactions [44].

Expression profiles of cellular proteins are also reported to change when cells are treated with CDs. A recent investigation studies the effects of HPβCD and HPγCD treatment on the fibroblasts derived from Niemann–Pick disease type C (NPC) patients. Pairwise comparison of the expression profiles of fibroblasts from NPC patients treated with HPβCD or HPγCD showed 137 and 170 proteins with more than twofold alteration in expression levels compared to untreated control. Interestingly, treatment with CDs also affected the expression level of the proteins that might be implicated in causing the disease. After treatment, the expression level of these proteins in patient-driven fibroblasts shifted toward values reported for healthy cells [55].

5 Cytotoxicity of CDs

While CDs have been useful in modulating membrane lipids in live cells, any application of CDs has to be performed with close attention to their potential cytotoxicity. This toxicity is in fact caused by the ability of CDs to extract lipids from cellular membranes, thereby disrupting membrane integrity. Cytotoxicity of CDs is primarily divided into two categories: changes in cell morphology and total cell disruption. CDs are reported to cause morphological changes and total hemolysis in human erythrocytes, with βCD being the most potent in this process [43, 56]. An investigation on the effects of MβCD and HPβCD on bovine pulmonary artery endothelial cells shows that at high concentrations, significant levels of cholesterol are removed from the cells, cell morphology is altered, membrane fatty acids composition is significantly changed, membrane proteins are altered, and overall cytotoxicity is recorded. A large difference between the alterations caused by the two derivatives led the investigators to suggest using HPβCD for cholesterol depletion studies as it does not cause cytotoxicity [57]. Other studies investigating the cytotoxicity of αCDs and βCDs separately on Caco-2 cells using proliferation assays also report the same potency profile. Although there are some discrepancies, trimethylated-αCD (TMαCD) and MβCD are the most potent of αCDs and βCDs tested on cells [58, 59]. A study on the cytotoxicity of CDs and their hydroxypropylated derivatives suggested that βCD is the most and HPαCD is the least potent of all six CDs in causing toxicity on human erythrocytes and P388 cells. These results indicate that the toxicity of the CDs is not cell specific, as the mechanism proposed for causing the toxicity is mentioned to be the removal of basic membrane components [60].

Extraction of proteins from the membrane is also mentioned as another reason for cytotoxicity. A study on effects of CDs on the membrane of erythrocytes mentions presence of ectoprotein, an integral transmembrane protein, in the aqueous solution after treatment with βCD. Although it is suggested that the protein removal is a consequential effect of CDs interacting with lipids, the extraction of proteins will adversely affect the cells regardless [43, 61]. This is also reported in another study on endothelial cells. The toxicity is mentioned to be caused by loss of cellular membrane fluidity through removing cholesterol and eventual leakage of cell's vital molecules including proteins [57].

6 Future Directions

Despite significant research performed on the application of CDs to modulate lipid composition in synthetic or cellular membranes, outstanding questions still exist. A detailed, molecular-level understanding of how CDs of different structures interact with phospholipids is still lacking. While it is known that CDs primarily interact with the acyl chains of phospholipids, it is not clear how CDs gain access to acyl chains in synthetic or biological membranes where the only part of the phospholipids that is available in an aqueous solution is the headgroup. While experimental studies have been useful in partially determining the structure of the CD–lipid inclusion complex and the kinetics of the interactions, a concerted effort from both experiments and simulations could help uncover key questions in this regard. In addition, a detailed understanding of the fate of proteins upon CD-mediated lipid exchange is missing. It is known that when lipids are extracted, proteins might get secreted by the cells as a result of loss of membrane integrity. However, despite a recent study [42], the effect of the lipid environment on protein localization, following CD-mediated lipid exchange, is less understood.

There is significant untapped potential for the use of CDs for mechanistic studies on the biophysical and biochemical properties of the cell plasma membrane. The use of CDs for the synthesis of asymmetric vesicles is still a relatively new field and will likely continue to be employed as a means to study membrane biophysical properties, interleaflet interactions, and protein localization. In addition, the use of CDs for lipid exchange in live mammalian [41, 42, 45, 53, 54] or non-mammalian cells [62] has opened new avenues for mechanistic studies on cell membrane biology. While such lipid-exchange studies will likely have to be optimized for each individual cell type, especially when it comes to the exchange of phospholipids, they provide many opportunities for probing membrane structure and function in live cells.

Acknowledgment

This work was supported by the NIH award R15ES030140 to AF.

References

1. Szejtli J (1998) Introduction and general overview of cyclodextrin chemistry. Chem Rev 98 (5):1743–1754
2. Bender ML, Komiyama M (2012) Cyclodextrin chemistry, vol 6. Springer Science & Business Media, Berlin
3. Davis ME, Brewster ME (2004) Cyclodextrin-based pharmaceutics: past, present and future. Nat Rev Drug Discov 3(12):1023–1035
4. Christian AE, Haynes MP, Phillips MC, Rothblat GH (1997) Use of cyclodextrins for manipulating cellular cholesterol content. J Lipid Res 38(11):2264–2272
5. Zidovetzki R, Levitan I (2007) Use of cyclodextrins to manipulate plasma membrane cholesterol content: evidence, misconceptions and control strategies. Biochim Biophys Acta 1768 (6):1311–1324
6. Pitha J, Irie T, Sklar PB, Nye JS (1988) Drug solubilizers to aid pharmacologists: amorphous cyclodextrin derivatives. Life Sci 43 (6):493–502
7. Jin Z-Y (2013) Cyclodextrin chemistry: preparation and application. World Scientific, Singapore
8. Shieh WJ, Hedges A (1996) Properties and applications of cyclodextrins. J Macromol Sci A Pure Appl Chem 33(5):673–683
9. Hammoud Z, Khreich N, Auezova L, Fourmentin S, Elaissari A, Greige-Gerges H (2019) Cyclodextrin-membrane interaction in drug delivery and membrane structure maintenance. Int J Pharm 564:59–76
10. Uekama K (2004) Design and evaluation of cyclodextrin-based drug formulation. Chem Pharm Bull (Tokyo) 52(8):900–915
11. Szente L, Fenyvesi E (2017) Cyclodextrin-lipid complexes: cavity size matters. Struct Chem 28 (2):479–492
12. Ravichandran R, Divakar S (1998) Inclusion of ring a of cholesterol inside the β-cyclodextrin cavity: evidence from oxidation reactions and structural studies. J Incl Phenom Mol Recognit Chem 30(3):253–270
13. Yu Y, Chipot C, Cai W, Shao X (2006) Molecular dynamics study of the inclusion of cholesterol into cyclodextrins. J Phys Chem B 110 (12):6372–6378
14. Loftsson T, Magnúsdóttir A, Másson M, Sigurjónsdóttir JF (2002) Self-association and cyclodextrin solubilization of drugs. J Pharm Sci 91 (11):2307–2316
15. López CA, de Vries AH, Marrink SJ (2011) Molecular mechanism of cyclodextrin mediated cholesterol extraction. PLoS Comput Biol 7(3):e1002020
16. Phillips MC, Johnson WJ, Rothblat GH (1987) Mechanisms and consequences of cellular cholesterol exchange and transfer. Biochim Biophys Acta 906(2):223–276
17. Denz M, Haralampiev I, Schiller S, Szente L, Herrmann A, Huster D, Müller P (2016) Interaction of fluorescent phospholipids with cyclodextrins. Chem Phys Lipids 194:37–48
18. Szejtli J, Cserhati T, Szögyi M (1986) Interactions between cyclodextrins and cell-membrane phospholipids. Carbohydr Polym 6 (1):35–49
19. Sanemasa I, Osajima T, Deguchi T (1990) Association of C5–C9 normal alkanes with cyclodextrins in aqueous medium. B Chem Soc Jpn 63(10):2814–2819
20. Tanhuanpää K, Cheng KH, Anttonen K, Virtanen JA, Somerharju P (2001) Characteristics of pyrene phospholipid/γ-cyclodextrin complex. Biophys J 81(3):1501–1510
21. Anderson TG, Tan A, Ganz P, Seelig J (2004) Calorimetric measurement of phospholipid interaction with methyl-β-cyclodextrin. Biochemistry 43(8):2251–2261
22. Nishijo J, Shiota S, Mazima K, Inoue Y, Mizuno H, Yoshida J (2000) Interactions of

cyclodextrins with dipalmitoyl, distearoyl, and dimyristoyl phosphatidyl choline liposomes. A study by leakage of carboxyfluorescein in inner aqueous phase of unilamellar liposomes. Chem Pharm Bull (Tokyo) 48(1):48–52

23. Fauvelle F, Debouzy JC, Crouzy S, Göschl M, Chapron Y (1997) Mechanism of α-cyclodextrin-Induced hemolysis. 1. The two-step extraction of phosphatidylinositol from the membrane. J Pharm Sci 86 (8):935–943
24. Debouzy JC, Fauvelle F, Crouzy S, Girault L, Chapron Y, Göschl M, Gadelle A (1998) Mechanism of α-cyclodextrin induced hemolysis. 2. A study of the factors controlling the association with serine-, ethanolamine-, and choline-phospholipids. J Pharm Sci 87 (1):59–66
25. Fauvelle F, Debouzy JC, Nardin R, Gadelle A (1994) Nuclear magnetic resonance study of a polar headgroup determined α-cyclodextrin-phospholipid association. Bioelectrochem Bioenerg 33(1):95–99
26. Coisne C, Tilloy S, Monflier E, Wils D, Fenart L, Gosselet F (2016) Cyclodextrins as emerging therapeutic tools in the treatment of cholesterol-associated vascular and neurodegenerative diseases. Molecules 21(12):E1748
27. Pike LJ (2006) Rafts defined: a report on the keystone symposium on lipid rafts and cell function. J Lipid Res 47(7):1597–1598
28. Kilsdonk EP, Yancey PG, Stoudt GW, Bangerter FW, Johnson WJ, Phillips MC, Rothblat GH (1995) Cellular cholesterol efflux mediated by cyclodextrins. J Biol Chem 270 (29):17250–17256
29. Yancey PG, Rodrigueza WV, Kilsdonk EP, Stoudt GW, Johnson WJ, Phillips MC, Rothblat GH (1996) Cellular cholesterol efflux mediated by cyclodextrins demonstration of kinetic pools and mechanism of efflux. J Biol Chem 271(27):16026–16034
30. Steck TL, Ye J, Lange Y (2002) Probing red cell membrane cholesterol movement with cyclodextrin. Biophys J 83(4):2118–2125
31. Rothblat GH, de la Llera-Moya M, Atger V, Kellner-Weibel G, Williams DL, Phillips MC (1999) Cell cholesterol efflux: integration of old and new observations provides new insights. J Lipid Res 40(5):781–796
32. Litz JP, Thakkar N, Portet T, Keller SL (2016) Depletion with cyclodextrin reveals two populations of cholesterol in model lipid membranes. Biophys J 110(3):635–645
33. Rouquette-Jazdanian AK, Pelassy C, Breittmayer JP, Aussel C (2006) Revaluation of the role of cholesterol in stabilizing rafts implicated in T cell receptor signaling. Cell Signal 18 (1):105–122
34. Hao M, Lin SX, Karylowski OJ, Wüstner D, McGraw TE, Maxfield FR (2002) Vesicular and non-vesicular sterol transport in living cells. The endocytic recycling compartment is a major sterol storage organelle. J Biol Chem 277(1):609–617
35. Foster LJ, De Hoog CL, Mann M (2003) Unbiased quantitative proteomics of lipid rafts reveals high specificity for signaling factors. Proc Natl Acad Sci USA 100(10):5813–5818
36. Simons K, Vaz WL (2004) Model systems, lipid rafts, and cell membranes. Annu Rev Biophys Biomol Struct 33:269–295
37. Scheiffele P, Rietveld A, Wilk T, Simons K (1999) Influenza viruses select ordered lipid domains during budding from the plasma membrane. J Biol Chem 274(4):2038–2044
38. Ikonen E (2001) Roles of lipid rafts in membrane transport. Curr Opin Cell Biol 13 (4):470–477
39. López CA, de Vries AH, Marrink SJ (2013) Computational microscopy of cyclodextrin mediated cholesterol extraction from lipid model membranes. Sci Rep 3:2071
40. Rodríguez-Acebes S, de la Cueva P, Fernández-Hernando C, Ferruelo AJ, Lasunción MA, Rawson RB, Martínez-Botas J, Gómez-Coronado D (2009) Desmosterol can replace cholesterol in sustaining cell proliferation and regulating the SREBP pathway in a sterol-Δ24-reductase-deficient cell line. Biochem J 420 (2):305–318
41. Kim JH, Singh A, Del Poeta M, Brown DA, London E (2017) The effect of sterol structure upon clathrin-mediated and clathrin-independent endocytosis. J Cell Sci 130 (16):2682–2695
42. Delle Bovi RJ, Kim J, Suresh P, London E, Miller WT (2019) Sterol structure dependence of insulin receptor and insulin-like growth factor 1 receptor activation. Biochim Biophys Acta Biomembr 1861(4):819–826
43. Ohtani Y, Irie T, Uekama K, Fukunaga K, Pitha J (1989) Differential effects of α-, β-and γ-cyclodextrins on human erythrocytes. Eur J Biochem 186(1–2):17–22
44. Motoyama K, Arima H, Toyodome H, Irie T, Hirayama F, Uekama K (2006) Effect of 2, 6-di-O-methyl-α-cyclodextrin on hemolysis and morphological change in rabbit's red blood cells. Eur J Pharm Sci 29(2):111–119
45. Tanhuanpää K, Somerharju P (1999) γ-Cyclodextrins greatly enhance translocation of hydrophobic fluorescent phospholipids from vesicles to cells in culture. Importance of

molecular hydrophobicity in phospholipid trafficking studies. J Biol Chem 274 (50):35359–35366

46. Huang Z, London E (2013) Effect of cyclodextrin and membrane lipid structure upon cyclodextrin–lipid interaction. Langmuir 29 (47):14631–14638
47. Lin Q, London E (2014) Preparation of artificial plasma membrane mimicking vesicles with lipid asymmetry. PLoS One 9(1):e87903
48. Clair JWS, London E (2019) Effect of sterol structure on ordered membrane domain (raft) stability in symmetric and asymmetric vesicles. Biochim Biophys Acta Biomembr 1861 (6):1112–1122
49. Cheng HT, London E (2009) Preparation and properties of asymmetric vesicles that mimic cell membranes effect upon lipid raft formation and transmembrane helix orientation. J Biol Chem 284(10):6079–6092
50. Son M, London E (2013) The dependence of lipid asymmetry upon phosphatidylcholine acyl chain structure. J Lipid Res 54(1):223–231
51. Son M, London E (2013) The dependence of lipid asymmetry upon polar headgroup structure. J Lipid Res 54(12):3385–3393
52. Kainu V, Hermansson M, Somerharju P (2008) Electrospray ionization mass spectrometry and exogenous heavy isotope-labeled lipid species provide detailed information on aminophospholipid acyl chain remodeling. J Biol Chem 283(6):3676–3687
53. Kainu V, Hermansson M, Somerharju P (2010) Introduction of phospholipids to cultured cells with cyclodextrin. J Lipid Res 51 (12):3533–3541
54. Li G, Kim J, Huang Z, St Clair JR, Brown DA, London E (2016) Efficient replacement of plasma membrane outer leaflet phospholipids and sphingolipids in cells with exogenous lipids. Proc Natl Acad Sci USA 113 (49):14025–14030
55. Szente L, Singhal A, Domokos A, Song B (2018) Cyclodextrins: assessing the impact of cavity size, occupancy, and substitutions on cytotoxicity and cholesterol homeostasis. Molecules 23(5):E1228
56. Irie T, Otagiri M, Sunada M, Uekama K, Ohtani Y, Yamada Y, Sugiyama Y (1982) Cyclodextrin-induced hemolysis and shape changes of human erythrocytes *in vitro*. J Pharmacobiodyn 5(9):741–744
57. Hinzey AH, Kline MA, Kotha SR, Sliman SM, Butler ES, Shelton AB, Gurney TR, Parinandi NL (2012) Choice of cyclodextrin for cellular cholesterol depletion for vascular endothelial cell lipid raft studies: cell membrane alterations, cytoskeletal reorganization and cytotoxicity. Indian J Biochem Biophys 49(5):329–341
58. Kiss T, Fenyvesi F, Bácskay I, Váradi J, Fenyvesi E, Iványi R, Szente L, Tósaki A, Vecsernyés M (2010) Evaluation of the cytotoxicity of β-cyclodextrin derivatives: evidence for the role of cholesterol extraction. Eur J Pharm Sci 40(4):376–380
59. Róka E, Ujhelyi Z, Deli M, Bocsik A, Fenyvesi É, Szente L, Fenyvesi F, Vecsernyés M, Váradi J, Fehér P, Gesztelyi R, Félix C, Perret F, Bácskay IK (2015) Evaluation of the cytotoxicity of α-cyclodextrin derivatives on the caco-2 cell line and human erythrocytes. Molecules 20(11):20269–20285
60. Leroy-Lechat F, Wouessidjewe D, Andreux J-P, Puisieux F, Duchêne D (1994) Evaluation of the cytotoxicity of cyclodextrins and hydroxypropylated derivatives. Int J Pharm 101 (1–2):97–103
61. Irie T, Uekama K (1997) Pharmaceutical applications of cyclodextrins. III Toxicological issues and safety evaluation. J Pharm Sci 86 (2):147–162
62. LaRocca TJ, Pathak P, Chiantia S, Toledo A, Silvius JR, Benach JL, London E (2013) Proving lipid rafts exist: membrane domains in the prokaryote Borrelia burgdorferi have the same properties as eukaryotic lipid rafts. PLoS Pathog 9(5):e1003353

Chapter 10

New Family of Fluorescent Probes for Characterizing Depth-Dependent Static and Dynamic Properties of Lipid/Water Interfaces

Moirangthem Kiran Singh, Him Shweta, and Sobhan Sen

Abstract

The important biological processes that help communicate between intra- and extracellular environments take place at cell membrane/water interface. However, molecular interactions at these interfaces are strongly affected by the depth-dependent hydration and local environmental polarity across the lipid/water interface and also the lipid ordering. A new family of membrane probes based on 4-aminophthalimide (4AP-Cn) have been synthesized, which are of particular interest here because of their extreme sensitivity towards sensing depth-dependent polarity, hydration and energy transfer dynamics at model lipid/water interfaces in sub-nanometre length scale. We envisage that these new probes will be extremely useful for characterizing the static and dynamic properties of naturally occurring membrane/water interfaces as well. This chapter presents the protocol for 4AP-Cn synthesis, as well as detailed experimental and MD simulation methods for measuring depth-dependent polarity, hydration and multi-molecular energy transfer dynamics at lipid/water interfaces of gel and fluid phases of lipid bilayer.

Keywords Lipid/water interface, Membrane probes, 4-Aminophthalimide, Polarity and hydration at interfaces, Lipid phases, Molecular dynamics simulations of lipid bilayers

1 Introduction

Biological membranes are complex two-dimensional assemblies of lipids, proteins and carbohydrates [1, 2]. Membranes are dynamic, which perform various molecular interactions for functioning of living organisms [2, 3]. However, molecular interactions at lipid membrane/water interface are strongly affected by the depth-dependent hydration state, local proton activity, hydrogen bond availability, hydration and its dynamics and the lipid ordering/packing [3–6]. Due to differential partitioning of water molecules near the hydrophilic heads and hydrophobic tails of membrane lipids, the lipid/water interfaces become highly anisotropic and disordered at sub-nanometre length scale [7–9]. As a result, properties such as local dielectric constant, fluidity and the extent of

Rajendra Prasad and Ashutosh Singh (eds.), *Analysis of Membrane Lipids*, Springer Protocols Handbooks,
https://doi.org/10.1007/978-1-0716-0631-5_10,

water penetration into bilayer change drastically in a depth-dependent manner [8–11]. Such changes can eventually influence many vital processes occurring at the membrane/water interfaces [12–15]. Hence, precise knowledge of depth-dependent polarity, hydration and molecular interactions at these interfaces is of vital importance, not only to explain the fundamental membrane functions [16], but also to develop efficient drug delivery, drug detoxification and artificial energy conversion systems [17–20].

Static and dynamic fluorescence techniques have extreme sensitivity, which can provide precise information of polarity, hydration, fluidity and dynamics at the molecular length scale. Thus, small fluorescent (reporter) molecules adsorbed at different depths of lipid/water interface can probe the varying solute–water and solute–lipid interactions, providing vital information about the local dielectric environment change in sub-nanometre length scale and also its effects on multi-molecular interactions within the lipid bilayer. Several fluorescence probes, such as NBD, BODIPY, AS, 3-hydroxyflavone derivatives, Patman, Laudran, tryptophan, etc., have been used to monitor the solute–solvent interactions and polarity change at lipid/water interfaces [21–29]. However, many of them were found to spontaneously translocate between membranes, and also the alkyl chains attached to many of such probes were found to bend back toward the membrane/water interface, perturbing the lipid packing [21–23]. Moreover, most of these probes were only suitable for studying fluid phase of lipid bilayers. These issues hindered the use of these probes for probing the static and dynamic properties of lipid/water interfaces at sub-nanometre length scale in fluid as well as in gel phases of lipid bilayer.

This chapter deals with a new family of homologous series of lipophilic fluorescent probes based on (solvatochromic) 4-aminophthalimde (4AP-Cn), which were synthesized by attaching single alkyl chain of different lengths to the fluorescent 4AP moiety (Fig. 1). Based on overall hydrophobicity (octanol/water partition coefficient—log *P*) of these probes, they can reside at different depths across the lipid/water interfaces and respond to small changes in their local environments and energy transfer efficiencies at sub-nanometre length scale (*see* Fig. 1c, d) [8, 9]. There are several advantages of using 4AP-Cn derivatives over other membrane probes (e.g., NBD, BODIPY, Laudran, AS, spin-lipids, etc.) for studying solute–solvent interactions, polarity and dynamics at lipid/water interfaces [30, 31]:

1. 4AP moiety shows large fluorescence Stokes shifts depending on subtle change in its surrounding dielectric environment.
2. 4AP has long fluorescence lifetime, good fluorescence quantum yield and no interfering excited-state processes.

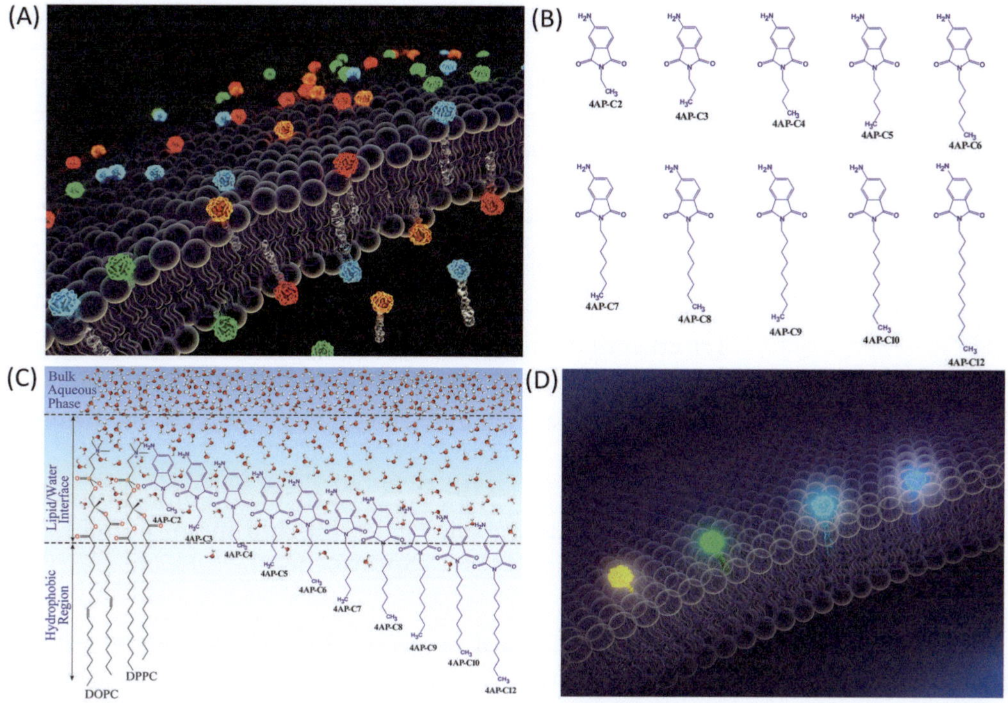

Fig. 1 (**a**) Cartoon representation of membrane bilayer, showing the probable positions of new class of 4-aminophthalimide based fluorescent probes. (**b**) Structure of designed and synthesized homologous series of 4AP-Cn ($n = 2$–10, 12) probes. (**c**) Schematic diagram showing the lipid/water interfacial region with possible (relative) locations of 4AP-Cn probes. Depending on hydrophobicity (log P), the probes are expected to be located at different depths across the lipid/water interfaces. (**d**) Schematic representation of 4AP-Cn resides at different depths across lipid/water interface, and subsequently, the probes would experience different local solvation to emit different colours of fluorescence. Figures (**c**) and (**d**) are reproduced from [8] with permission from the PCCP Owner Societies

3. It has molecular properties for probing dipolar as well as hydrogen bond interactions (H-bond donating and accepting) with surrounding molecules through its fluorescence spectral shifts.
4. Due to their small size and molecular structure, 4AP-Cn probes barely perturb the lipid packing when compared to other probes.
5. Because of the precise balance of hydrophilic and hydrophobic interactions, the fluorescent part (4AP moiety) of 4AP-Cn probes resides near the lipid/water interface.

The only disadvantage with 4AP-Cn probes is that they have low extinction coefficient (~3000–7000 M^{-1} cm^{-1}, depending on local environment). Thus, one would need to work with high concentrations to obtain good fluorescence signals from these probes.

In this chapter, we present the protocol for 4AP-Cn synthesis, as well as the detailed experimental and MD simulation methods for measuring local environmental polarity, hydration and multi-chromophoric energy transfer dynamics at lipid/water interfaces of gel phase and fluid phase of lipid bilayers[8, 9].

2 Materials

2.1 Synthesis of 4AP-Cn

- 4-aminophthalimide (4AP, TCI Chemicals), recrystallized from 1:1 ethanol/water mixture prior to synthesis of 4AP-Cn probes (store recrystallized 4AP at 4 °C and protect from light).
- Alkyl bromide (chains with 2–10, 12 carbons from TCI Chemicals).
- Dry acetonitrile (HPLC, Spectrochem), anhydrous K_2CO_3 (Fischer Scientific), silica gel (100–200 mesh, Merck), hexane and ethylacetate (HPLC, Spectrochem).

2.2 Small Unilamellar Vesicles with 4AP-Cn and Rhodamine-6G

- DOPC (1,2-dioleoyl-*sn*-glycero-3-phosphocholine) and DPPC (1,2-dipalmitoyl-*sn*-glycerol3-phosphocholine) lyophilized power, received from Avanti Polar Lipids and stored at −20 °C.
- Water (HPLC, Merck), 1 mM 4AP-Cn stock in methanol (stored at 4 °C), Rhodamine 6G (Rh6G, Sigma-Aldrich) and freezing mixture of 1:3 ratio of $CaCl_2$/ice.

2.3 Partition Coefficients (logP) of 4AP-Cn and Polarity Scale

- Octanol (HPLC, Spectrochem), water (HPLC, Merck) and 4AP-Cn ($n = 2$–10, 12).
- Methanol solution of 3–5 mM 4AP-C3; water, 2,2,2-trifluoroethanol, methanol, diethylene glycol, ethanol, butanol, pentanol, hexanol, decanol, 2-pentanol, 3-pentanol, tert-butanol, 2-ethyl-2-butanol and 3-ethyl-3-pentanol. All solvents from either Merck or Sigma-Aldrich or Spectrochem.

2.4 Relative Depth and Absolute Positions of Probes at Interfaces

- Thioacetamide (Spectrochem), water, 4AP-Cn/DPPC (1:35) and 4AP-Cn/DOPC (1:40) in form of small unilamellar vesicles (SUVs).
- 10 mM solution of 1-palmitoyl-2-(6,7-dibromostearoyl), phosphatidylcholine (6,7-Br2PC), 1-palmitoyl-2-(9,10-dibromostearoyl), phosphatidylcholine (9,10-Br2PC) in methanol, 100 mM DOPC and DPPC in methanol, water, 1 mM stock solutions of 4AP-Cn in methanol and freezing mixture—1:3 ratio of $CaCl_2$/ice. Bromo-lipids were from Avanti Polar Lipids.

2.5 Instruments

- Hot plate cum magnetic stirrer, teflon-coated magnetic stirring bar (Remi), rotary vacuum evaporator (Heidolph), ultrasonic bath sonicator (Praveen Scientific), vortex mixer (Remi), vacuum desiccator (Tarson) and suction pump (Tarson, Rocker 400).
- NMR Spectrometer (500 MHz, Bruker Advance III), LCMS (Waters SYNAPT G2), TEM (Jeol2100F), UV-Vis spectrophotometer (Shimadzu-2450) and Fluorescence spectrometer (Varian Cary Eclipse) with 3 × 3 mm quartz cuvettes (Starna Cells) and Time-correlated single photon counting (TCSPC) setup (FL920, Edinburgh Instruments, UK).

3 Methods

3.1 Experimental Methods

All experimental procedures were carried out in room temperature unless otherwise specified.

3.1.1 Synthesis of 4AP-Cn

- Take 162 mg (1 equiv.) of 4AP into 50 ml round-bottomed flask and then add 138 mg (1 equiv.) of K_2CO_3, 5 ml anhydrous acetonitrile and a magnetic stir bar into flask (*see* **Note 1**).
- Fix a refluxing condenser with flask and place the flask into oil bath. Set the temperature of oil bath within 65–70 °C. Turn on circulating chilled water for condenser to control refluxing of acetonitrile.
- Add 67 μl of *n*-ethylbromide (0.9 equiv.) drop-wise into the reaction mixture at regular intervals and continue refluxing for 24 h. The reaction will follow according to Scheme 1. After 24 h, stop the reaction and cool the mixture to room temperature (*see* **Note 2**). Evaporate solvents from reaction mixture using rotary vacuum evaporator to obtain residual solid (*see* **Note 3**).
- Prepare silica gel slurry for column chromatography in 2:1 hexane/ethylacetate mixture and ensure that there is sufficient solvent on top of silica layer.
- Dissolve dried crude product into 2–4 ml of ethylacetate and put the soluble fraction slowly on to the top of column. Carry out elution using 2:1 (v/v) hexane/ethylacetate eluent. Collect the desired parts of elutes in small fractions in conical flasks and analyse their contents by thin layer chromatography (TLC). Concentrate eluent to dryness under reduced pressure at ≤40 °C using rotary vacuum evaporator. Then, dry out solid product overnight in vacuum desiccator attached to suction pump through a low temperature trap and recrystallize the product from 1:1 (v/v) water/ethanol mixture to get final 4AP-Cn molecule.

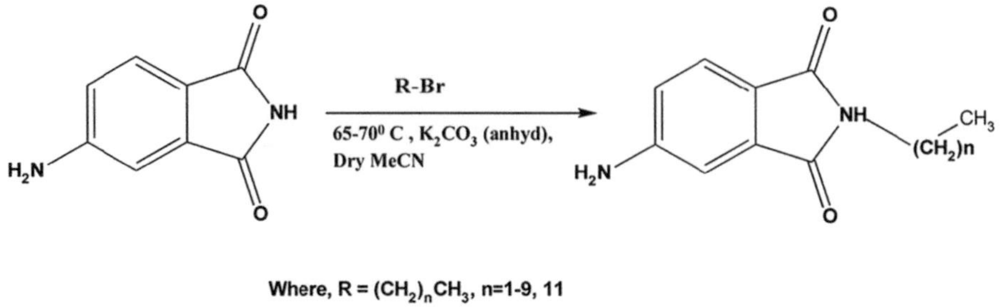

Scheme 1 Synthesis scheme of 4AP-Cn probes

- Dissolve final products in chloroform-d_3 to characterize product in ^{1}H NMR and also in mass spectrometry (*see* **Note 4**).
- Use above procedure to synthesize all 4AP-Cn ($n = 3$–10, 12) probes. The volume fractions of the eluent mixtures for column chromatography should be changed for purification of different 4AP-Cn probes of different hydrophobicity.

3.1.2 Preparation of SUVs with Probes

- Dissolve 100 mg of DOPC and DPPC in 1272 μl and 1362 μl of 1:1 (v/v) chloroform/methanol mixture in 5 ml glass tubes, respectively, to make 100 mM stock solutions of lipids.
- Mix 8.75 μl of stock DOPC or DPPC solution with 25 μl of 4AP-Cn stock methanol solution in 5 ml glass tube to make dye/lipid ratio of 1:35 (*see* **Note 5**). Evaporate residual solvent by gently warming. Finally dry up residual organic solvent in a vacuum desiccator connected to suction pump through a trap.
- Rehydrate dried lipid/4AP-Cn film in 500 μl water and let it stand for 2 h. Mix the solution in a vortex mixture till a milky suspension is obtained.
- Set up ultrasonic bath sonicator at 60 °C (*see* **Note 6**) and prepare freezing mixture of 1:3 of $CaCl_2$/ice (typical temperature is ~−20 °C).
- Put the glass bottle containing lipid/4AP-Cn into freezing mixture for 5 min and then immediately transfer to bath sonicator for 10 min, followed by immediate return back into freezing mixture. Repeat this freeze-thaw-sonication cycle until the milky solution becomes clear (*see* **Note 7**).
- Confirm the formation and size of SUVs using TEM. Typical size (diameter) of SUVs prepared by this method was estimated to vary within ~50–150 nm.

3.1.3 Estimation of Partition Coefficients (logP) of Probes

- Dissolve known amount of 4AP-Cn in 1 ml octanol. Add this solution to 100 ml of octanol/water (1:1, v/v) in 500 ml conical flask with a stirring bid and stir the mixture for 2 days at room temperature.

- Transfer the mixture to 250 ml separating flask and keep it in vibration isolation to separate the phase spontaneously over 2–3 days.
- Measure concentrations of 4AP-Cn in each phase (octanol and water) using UV-Vis spectrophotometer and taking extinction coefficient of 4AP-Cn in octanol and water as 7835 M^{-1} cm^{-1} and 3382 M^{-1} cm^{-1} at 378 nm, respectively.
- Calculate log P value from known concentrations of 4AP-Cn in each phase as,

$$\log P = \log \frac{[4\text{AP}-\text{Cn}]_{\text{octanol}}}{[4\text{AP}-\text{Cn}]_{\text{water}}} \quad (1)$$

3.1.4 Steady-State Fluorescence Spectra and Quantification of Polarity at Lipid/Water Interface

- Add 200 μl of vesicle/dye solution (4AP-Cn/lipid = 1:35) into a pre-cleaned 3 × 3 mm quartz cuvette and record the steady-state fluorescence spectra. For this measurement, set the excitation at 375 nm and the band pass for excitation and emission to 5 and 1.5 nm, respectively.
- Perform instrumental correction on measured spectra using standard spectra of quinine sulphate to obtain final emission spectra in photons cm^{-1} vs. cm^{-1}.
- Figure 2 shows the corrected spectra of 4AP-Cn probes at the lipid/water interfaces of gel-phase DPPC and fluid-phase DOPC vesicles at room temperature. Figure 2a shows spectral shifts of 4AP-Cn probes at DPPC/water interface toward higher wavenumber in the sequence: (4AP-C3 ≈ -C4) < (4AP-C5 ≈- -C6) < (4AP-C7 ≈ -C8) < (4AP-C9 ≈ -C10), but with reverse trend for 4AP-C12. However, Fig. 2b shows only subtle but monotonic shift of spectra at fluid-phase DOPC/water interface toward higher wave number from 4AP-C2 to 4AP-C9 with reverse trend for 4AP-C12 (*see* **Note 8**).
- To create the polarity (E_T^N) scale, measure fluorescence spectra of 1–5 mM 4AP-C3 in water and alcohols of different polarities, such as 2,2,2-trifluoroethanol, methanol, diethylene glycol, ethanol, butanol, pentanol, hexanol, decanol, 2-pentanol, 3-pentanol, tert-butanol, 2-ethyl-2-butanol and 3-ethyl-3-pentanol at the same instrumental condition used above.
- Correct all spectra for instrumental errors using quinine sulphate and plot the fluorescence peak frequencies of 4AP-C3 in 15 different solvents against E_T^N values of corresponding solvents [32]. This will create a linear E_T^N polarity scale from which the depth-dependent polarities of 4AP-Cn at the DPPC/water and DOPC/water interfaces can be estimated.

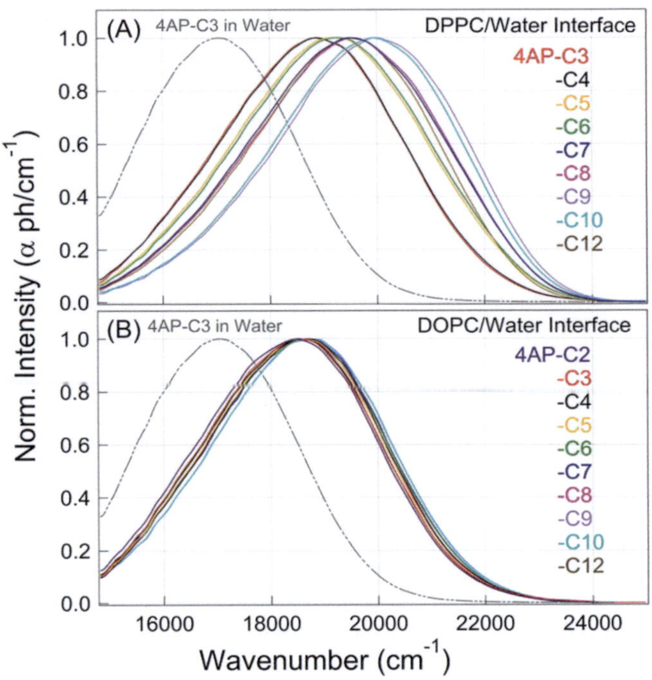

Fig. 2 Corrected fluorescence spectra of 4AP-Cn at (**a**) gel-phase DPPC/water interface and (**b**) fluid-phase DOPC/water interface. Plots show the (control) spectrum of 4AP-C3 in water. Data show stepwise shift at DPPC/water interface, while subtle but continuous shift at DOPC/water interface. Reproduced from [8] with permission from the PCCP Owner Societies

- Figure 3a shows steeply decreasing E_T^N value from 0.53 to 0.18 at gel-phase DPPC/water interface with increasing log *P* of 4AP-Cn probes in stepwise manner, but a reverse trend is observed for 4AP-C12. Figure 3b, however, shows only subtle but continuous decrease of E_T^N value from 0.64 to 0.54 with increasing log *P* at fluid-phase DOPC/water interface, with opposite trend for 4AP-C12.

3.1.5 Relative Depth Variation of Probes

- The relative depth variation of 4AP-Cn probes at lipid/water interface can be measured by monitoring the relative fluorescence quenching of the probes by a water-soluble quencher, such as thioacetamide, which resides in the water-rich region near the lipid head groups.
- Make 2.1 M Thioacetamide stock solution by dissolving 789 mg thioacetamide in 5 ml water.
- Measure fluorescence spectra of 4AP-Cn probes in lipid vesicles with sequentially adding thioacetamide from stock.

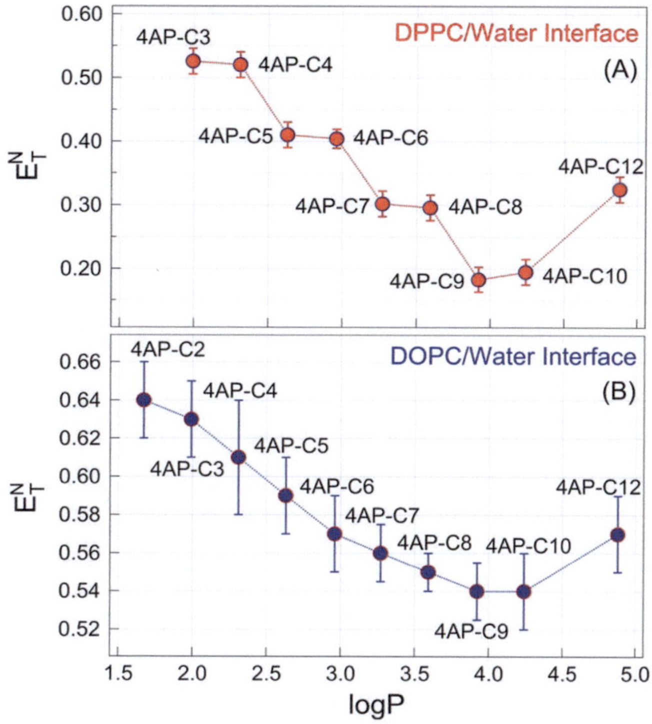

Fig. 3 Plots of E_T^N variation with log *P* at (**a**) gel-phase DPPC/water interface and (**b**) fluid-phase DOPC/water interface. Data show stepwise and continuous changes at the DPPC/water and DOPC/water interfaces, respectively. Error bars (±SD) are obtained from triplicate measurements. Reproduced from [8] with permission from the PCCP Owner Societies

- Using emission maxima of above spectra, create Stern–Volmer plot of each of 4AP-Cn probes in DPPC and DOPC vesicles. Figure 4a shows that relative quenching of probes follows stepwise behaviour at DPPC/water interface as: (4AP-C3≈-C4)>(4AP-C5≈-C6)>(4AP-C7≈-C8≈-C12)>(4-AP - C9 ≈ - C10). However, nearly continuous change in relative quenching of probes is observed at DOPC/water interface, with 4AP-C12 showing the reverse trend. These results indicate that depending on the overall hydrophobicity of 4AP-Cn probes, the relative depths of these probes can vary across the lipid/water interfaces at sub-nanometre length scale, which eventually report the relative variation of depth-dependent local polarity at the interfaces (Fig. 3).

3.1.6 Absolute Position of Probes at Interface

- The absolute position of 4AP-Cn probes can be measured using the parallax method [21, 22]. For this purpose use two bromo-labelled-PC quenchers, viz. 6,7-Br2PC and 9,10-Br2PC.

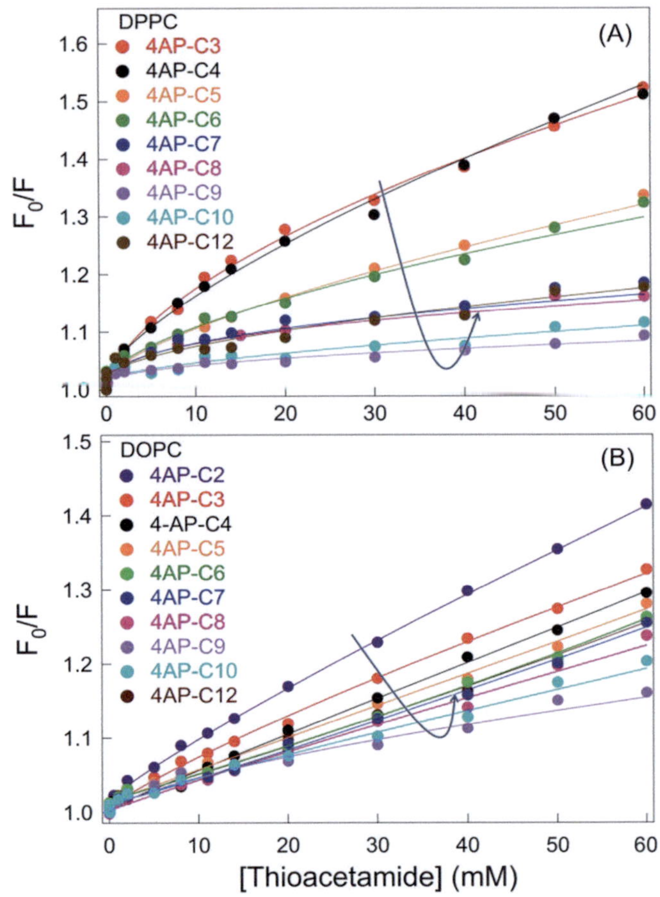

Fig. 4 Stern–Volmer plots of fluorescence quenching of 4AP-Cn probes at different concentrations of thioacetamide in (**a**) DPPC and (**b**) DOPC vesicles. Here, F_0 and F are the fluorescence intensities in absence and presence of quencher, respectively. Reproduced from [8] with permission from the PCCP Owner Societies

- In an aliquot, mix 1.31 μl of 6,7-Br2PC with 7.34 μl of lipid (100 mM DOPC) in a 5 ml glass bottle with bromo-PC/lipid ratio of 15:85. Add 25 μl of stock 4AP-Cn to the mixture and prepare the lipid vesicles using procedure mentioned in Subheading 3.1.2.
- In other set of sample, mix 1.31 μl 9,10-Br2PC with 7.34 μl of lipid (100 mM DOPC) with bromo-PC/lipid ratio of 15:85 in a 5 ml glass bottle. Add 25 μl of 4AP-Cn to this mixture and prepare the lipid vesicle as above.
- Take 200 μl of DOPC/6,7-Br2PC/4AP-Cn mixture solution into 3 × 3 mm quartz cuvette and record the steady-state fluorescence spectra with same instrumental conditions. Repeat these steps for DOPC/9,10-Br2PC/4AP-Cn systems as well.

- Calculated the distance of probes from the centre of bilayer using the parallax equation [21, 22]

$$Z_{cB} = \left[\frac{\left(\frac{-1}{\pi C}\right) ln \left(\frac{F_1}{F_2}\right) - L_{21}^2}{2L_{21}} \right] + Z_{c1} \quad (2)$$

where, C is concentration of quencher (bromo-PC) per unit area, F_1 is the fluorescence intensity of 4AP-Cn in presence of shallow quencher (6,7-Br_2PC) and F_2 is that in presence of equivalent mole of deep quencher (9,10-Br_2PC). L_{21} is the depth difference between shallow and deep quenchers (2.5 Å) and Z_{c1} is distance of shallow quencher from bilayer centre (10.8 Å for 6,7-Br_2PC). Calculate C by dividing the mole fraction of quencher lipid (in total DOPC lipid) by the surface area per lipid (70 Å^2 in case of DOPC) [21, 22].

- *No reliable parallax data in gel-phase DPPC vesicle could be obtained as the bromo-PCs get excluded out of the rigid gel-phase lipid bilayer and/or perturb the lipid packing.*

3.1.7 Multi-chromophoric Förster Resonance Energy Transfer at Interface

- It is possible to use 4AP-Cn probes for studying depth-dependent Förster Resonance Energy Transfer (FRET) at the lipid/water interfaces. For this purpose, 4AP-Cn can be used as *donor* and another probe molecule (e.g., Rhodamine-6G, Rh6G) as the energy *acceptor* as long as they come close to each other and there is sufficient overlap between the emission of 4AP-Cn probes and absorption of Rh6G (Fig. 5). Indeed, these *donor–acceptor* pairs show good overlap between their emission and absorption [9].
- The best way to monitor multi-chromophoric FRET at lipid/water interfaces is to measure the *donor* (4AP-Cn) fluorescence decays in the presence of different concentrations of acceptor (Rh6G) using picosecond-resolved time-correlated single photon counting (TCSPC) technique.
- Such measurements provide information of energy transfer efficiencies and closest approach between *donor* and *acceptor* at the lipid/water interface (*see* Fig. 5).
- To monitor the FRET process, measure donor (4AP-Cn) decays near its emission peak in the absence and presence of varying concentrations of acceptor (Rh6G) in the gel-phase DPPC/water and fluid-phase DOPC/water interfaces. For this purpose, the donor was excited at 375 nm using a picosecond diode laser (Edinburgh Instrument; pulse width ~60 ps) in TCSPC setup with instrument response function (IRF) of ~100 ps (*see* **Note 9**).

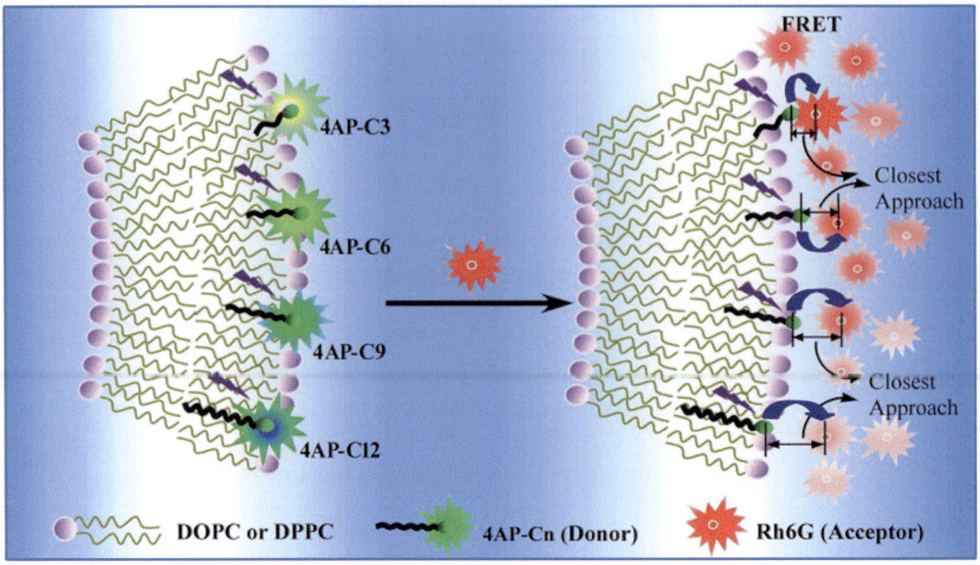

Fig. 5 Schematic representation of multi-chromophoric FRET process between 4AP-Cn donors and Rh6G acceptors at lipid/water interface. Reproduced from Sing et al. [9] with permission from the PCCP Owner Societies

- Set the final concentration of donor (4AP-Cn) in lipid vesicles to 50 μM. Vary the concentration of acceptor (Rh6G) from 5 to 300 μM in gel-phase DPPC and from 5 to 50 μM in fluid-phase DOPC vesicles. (Note that due to hydrophilic nature of the acceptor (Rh6G), it stays in the water-rich region near the lipid head-groups.)
- Figure 6 shows the fluorescence decays of 4AP-Cn donors in presence of varied concentrations of acceptor for two *donor–acceptor* pairs at the DPPC/water and DOPC/water interfaces (*see* [9] for full dataset). As can be seen, the fluorescence decays get faster on successive addition of acceptor (Rh6G) in the aqueous solution of lipid vesicles. It can be seen that the 4AP-Cn (donor) decays get saturated when concentration ratios of 4AP-Cn/Rh6G become nearly 1:1 at the DOPC/water interface. However, at gel-phase DPPC/water interface the decays of donor saturate at donor/acceptor concentration ratios which follow a stepwise behaviour, similar to that observed in polarity data in Fig. 3a.
- For uniform distribution of 4AP-Cn donors and Rh6G acceptors at the interface, a donor molecule can be surrounded by many acceptor molecules, as long as the acceptor concentration remains higher than the donor concentration. In such case, fluorescence decay of donor depends on acceptor concentration, lipid-bilayer fluidity, topology and local dielectric environment. For such multi-chromophoric FRET between donors and

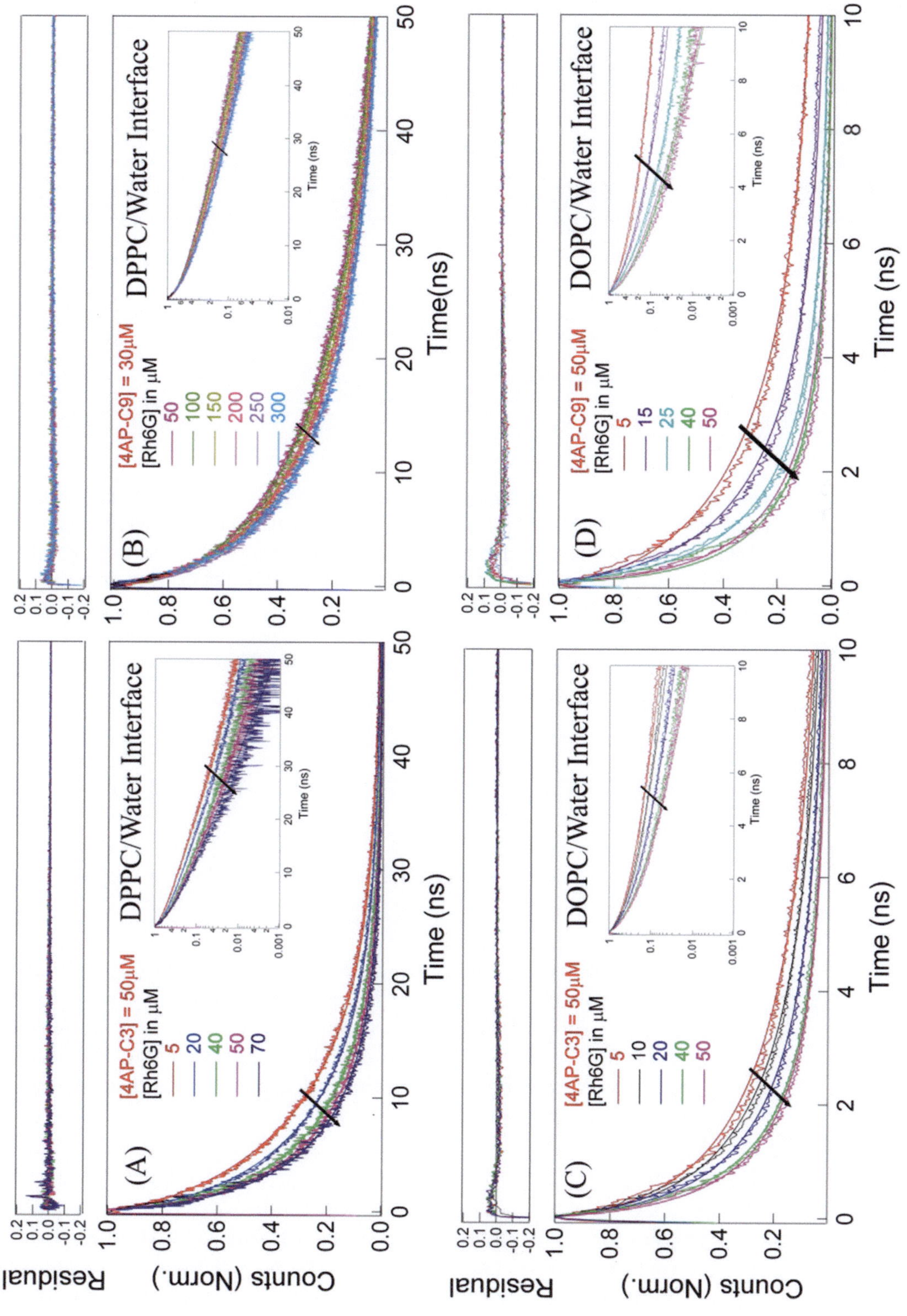

Fig. 6 Fluorescence decays of donor (4AP-C3 and 4AP-C9) at DPPC/water (panels **a** and **b**) and DOPC/water (panels **c** and **d**) interfaces in presence of varying concentration of acceptor (Rh6G). Global fits to data (Eq. 3) are shown as solid lines through raw data. Reproduced from [9] with permission from the PCCP Owner Societies

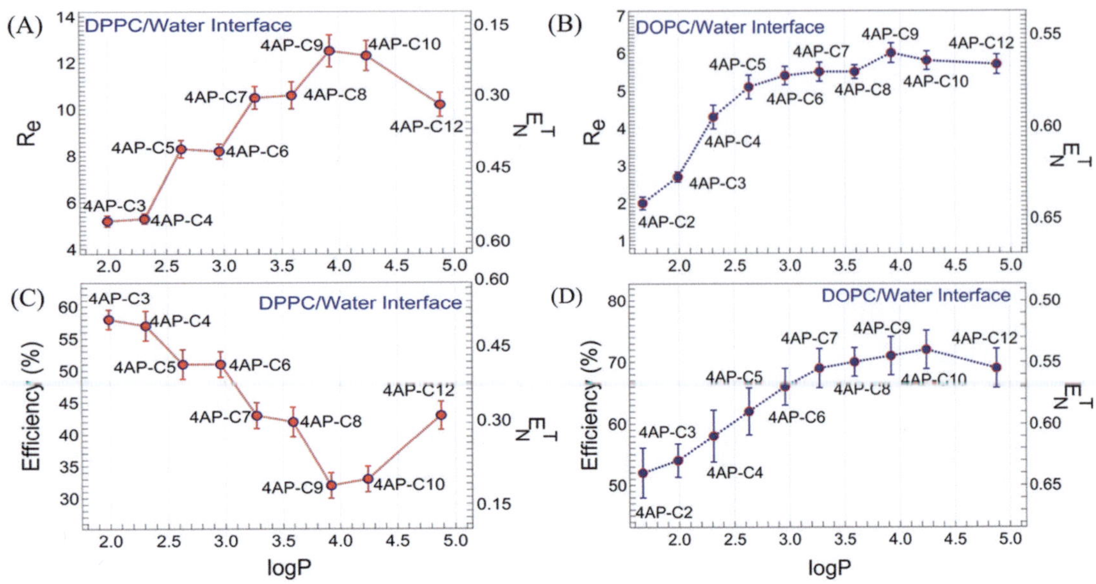

Fig. 7 Change of closest approach (R_e) and FRET efficiencies (E) between donors (4AP-Cn) and (multiple) acceptors (Rh6G) with donors' log P values at DPPC/water (**a** and **b**) and DOPC/water (**c** and **d**) interfaces. These plots also include the variation of corresponding E_N^T sensed by donors with that of donors' log P, which show the effect of local dielectric environment on the multi-chromophoric FRET. Reproduced from [9] with permission from the PCCP Owner Societies

acceptors, the acceptor-concentration-dependent donor decays are analysed globally using equation [33],

$$I_{DA}(t) = \exp\left(-\frac{t}{\overline{\tau}_D}\right) \times \exp\left\{-R_0^2 N \gamma\left[\frac{2}{3}, \left(\frac{R_0}{R_e}\right)^6 \left(\frac{t}{\overline{\tau}_D}\right)\right] \left(\frac{t}{\overline{\tau}_D}\right)^{d/6}\right\} \times \exp\left\{R_e^2 N\left(1 - \exp\left[-\left(\frac{R_0}{R_e}\right)^6 \left(\frac{t}{\overline{\tau}_D}\right)\right]\right)\right\} \quad (3)$$

This equation incorporates the closest distance (R_e) between donor (D) and acceptor (A). The $\gamma[..]$ is incomplete gamma function, d is dimensionality of acceptor distribution, N is numerical concentration of acceptor (molecules/unit area) and $\overline{\tau}_D$ is the intensity-averaged lifetime of donor decay in the absence of acceptor, which is given by [34]:

$$\overline{\tau}_D = \frac{a_1\tau_1^2 + a_2\tau_2^2}{a_1\tau_1 + a_2\tau_2} \quad (4)$$

- For global fitting of a dataset, make N free while keeping R_e and d as free *global* parameters. Use the pre-calculated R_0 from the steady-state fluorescence data and $\overline{\tau}_D$ from donor decay in absence of acceptor as fixed parameters in this fitting analysis [9].

- For measuring FRET efficiency (E), use the average lifetimes of 4AP-Cn donors in the absence and presence of *highest* concentration of acceptor (Rh6G) as [34],

$$E = 1 - \frac{\int I_{DA}(t)\mathrm{d}t}{\int I_D(t)\mathrm{d}t} = 1 - \frac{\langle \tau_{DA} \rangle}{\langle \tau_D \rangle} = 1 - \frac{a_1^{DA}\tau_1^{DA} + a_2^{DA}\tau_2^{DA}}{a_1^D\tau_1^D + a_2^D\tau_2^D} \quad (5)$$

where, $\langle \tau_D \rangle$and $\langle \tau_{DA} \rangle$are amplitude-weighted lifetimes of donors in the absence and presence of highest concentration of acceptor, respectively.

- Figure 6 shows the global fits to raw fluorescence decays. Global fit provides a single R_e value for a particular D-A pair for a given lipid/water interface.
- Very interestingly it can be seen that R_e and the FRET efficiency (E) for the D-A systems changes in stepwise manner with donor log P at gel-phase DPPC/water interface along with a reverse trend for 4AP-C12/Rh6G pair (Fig. 7a, c), similar to the stepwise variation of polarity $\left(E_N^T\right)$. On the other hand, R_e and E change continuously with donor log P—a feature similar as E_N^T variations with log P at fluid-phase DOPC/water interface (Fig. 7b, d). This data set again confirms that differential partitioning of solvatochromic 4AP-Cn donors and Rh6G acceptor at gel-phase DPPC/water and fluid-phase DOPC/water interfaces directly control the multi-chromophoric FRET at the interfaces, which are inherently influenced by the fine tuning of donors' (and acceptor's) local dielectric environment as well as the fluidity/rigidity of the lipid/water interfaces.

3.2 Simulation Methods

MD simulations provide a clear picture of biomolecular interactions at molecular level. Results obtained from MD simulation can be directly compared with experimental results. A detailed protocol is presented here for simulating 4AP-Cn and Rh6G probes at gel-phase DPPC/water and fluid-phase DOPC/water interfaces.

3.2.1 Starting Structures for MD Simulation

- *Coordinates of Probes*: Obtain initial coordinates in pdb format for 4AP-Cn and Rh6G by drawing these molecular structures in ChemDraw software. Optimize the energy of probes using Gaussian 09 software [35] at Hartree-Fock (HF) level using 6-31G* basis set. Generate "pdb" structure of the energy optimized probe using GaussView.
- *Excited State Charges of Probes*: Generate the electrostatic potential (ESP) on optimized 4AP-Cn structure at CIS/6-31G* level using Gaussian 09. Then calculate the excited state atomic partial charges of 4AP-Cn probes through two stage restrained electrostatic potential (RESP) charge methodology [36]. In this RESP methodology, first optimize charge by applying a hyperbolic restraint of 0.0005 on non-hydrogen atoms, keeping

other atoms free. Then at second stage, re-optimize and refit the charges using a hyperbolic restraint of 0.001 on non-hydrogen atom. At this stage freeze charges on all atoms except for methylene (CH_2) and methyl group (CH_3) of the carbon chains of 4AP-Cn probes. Use similar RESP method for obtaining ground-state partial charges of Rh6G atoms.

- *Force-Field Parameters of Probes*: Generate force-field parameters and topologies for 4AP-Cn and Rh6G probes using general amber force field (GAFF) [37] in antechamber module of AMBER-12 [38]. Convert force-field parameters and topology of probes into GROMACS format using python based ACPYPE (AnteChamber Python Parser Interface) script [39]. This is needed to run the simulation in GROMACS software package [40].
- *Coordinates for Lipid Bilayer*: Pre-equilibrated fluid-phase DOPC and gel-phase DPPC bilayer structures using Slipids (Stockholm lipid) force field [41] can be obtained from previous study [52]. The bilayer structures are formed by 64 lipid molecules in each leaflet, making a system of 128 lipid molecules. DOPC and DPPC bilayers contain 5120 and 3480 TIP3P [42] water molecules, respectively. These coordinates can be used further to generate "itp" (include topology) files for both DOPC and DPPC bilayers. Use Slipids (Stockholm lipid) force field to prepare topology files for DOPC and DPPC bilayer using "pdb2gmx"engine of GROMACS package for further run (*see* **Note 10**).
- *Incorporation of Probes into Lipid Bilayer*: Place the 4AP-Cn molecules near the lipid/water interface of DOPC and DPPC bilayers such that the carbon chains of the probes are aligned parallel to the carbon tail of the lipid bilayer. Use "*editconf*" engine of GROMACS with "translate" and "align" commands to place the probe at desired position. Use any visualization program to make sure that the carbon tails of probes are aligned parallel to the lipid hydrocarbon tail and the 4AP moiety is placed near the lipid/water interface. Prepare the topology files with force-field parameters for (TIP3P) water, 4AP-Cn, and lipid bilayer in a single input topology file using GROMACS. Prepare the Rh6G/lipid bilayer system using same procedure.

3.2.2 Molecular Simulation Protocol

- *Minimization*: Relax the bilayer/probe systems through energy minimization using steepest decent algorithm, so as to remove any bad contacts in the system. Restrain the position of lipids and 4AP-Cn (or Rh6G) probe with force constant of 1000 kJ mol^{-1} nm^{-2}. Next, minimize the system with reduced restraint of 500 kJ mol^{-1} nm^{-2} on probes. Finally, carry out minimization without restraint. Use CPU codes of GROMACS to perform these steps.

- *Equilibration*: Heat up the probe/bilayer system to production temperature in NVT ensemble for 100 ps using Berendsen thermostat [43]. During this process restrain the lipids and probe with a force constant of 1000 kJ mol^{-1} nm^{-2}. To maintain the fluid (liquid crystalline—L_α) phase of DOPC bilayer, heat up to temperature of 303 K. For (tilted) $L_{\beta'}$ gel phase of DPPC bilayer, heat up the system to 293 K. Make sure that probes and lipid bilayer are separately coupled to the heat bath through velocity rescale Berendsen thermostat. Next, equilibrate the system in NVT ensemble for 100 ps with decreased restraint of 500 kJ mol^{-1} nm^{-2} on probes and lipids. Run 100 ps restraint free equilibration after that. Now, equilibrate the system in NPT ensemble for 1 ns, restraining the lipids and probes with a force constant of 1000 kJ mol^{-1} nm^{-2}, followed by equilibration of 1 ns with reduced restraint of 500 kJ mol^{-1} nm^{-2} using Nose-Hoover thermostat [44, 45]. Finally, perform restraint free equilibration of 1 ns in NPT ensemble. During equilibration use semi-isotropic pressure coupling for lipid bilayers and LINCS algorithm [46] to restraint bonds. Use CPU codes of GROMACS to perform equilibration.
- *Production Simulation*: Perform 2 ns simulation in NPT ensemble using CPU codes prior to the final production run. Perform the production simulation in GPU (In our case, Nvidia Tesla K20 card) using Slipids (Stockholm lipid) force field for lipids, TIP3P for water, GAFF for 4AP-Cn, and Rh6G in GROMACS v5.0.5. Use time step of 2 fs for integration of MD and the particle mesh Ewald (PME) sum [47] to treat long-range electrostatic interaction with a real space cut off of 10 Å. Maintain 1 bar pressure using Parrinello-Rahmanbarostat [48] with an isothermal compressibility of 4.5×10^{-5} bar^{-1} and a coupling constant of 10 ps. Save the simulation snapshots at every 5 ps for further analysis.

3.2.3 Simulation Results of Lipid/Probe Systems

- Total eight lipid/4AP-Cn systems were simulated for 300 ns each in the fluid-phase DOPC and gel-phase DPPC bilayers, which includes four 4AP-Cn probes of different chain lengths (4AP-C3, -C6, -C9 and -C12) in DOPC and four 4AP-Cn probes (4AP-C3, -C7, -C9 and -C12) in DPPC bilayers (total simulation run of 2.4 μs). For Rh6G/lipid systems, separate 25 ns simulations were performed in DPPC and DOPC lipid bilayer.
- Position and angle distributions of these probes in fluid and gel phases of bilayers can be calculated to explore the importance of hydration in understanding the local dielectric environment at

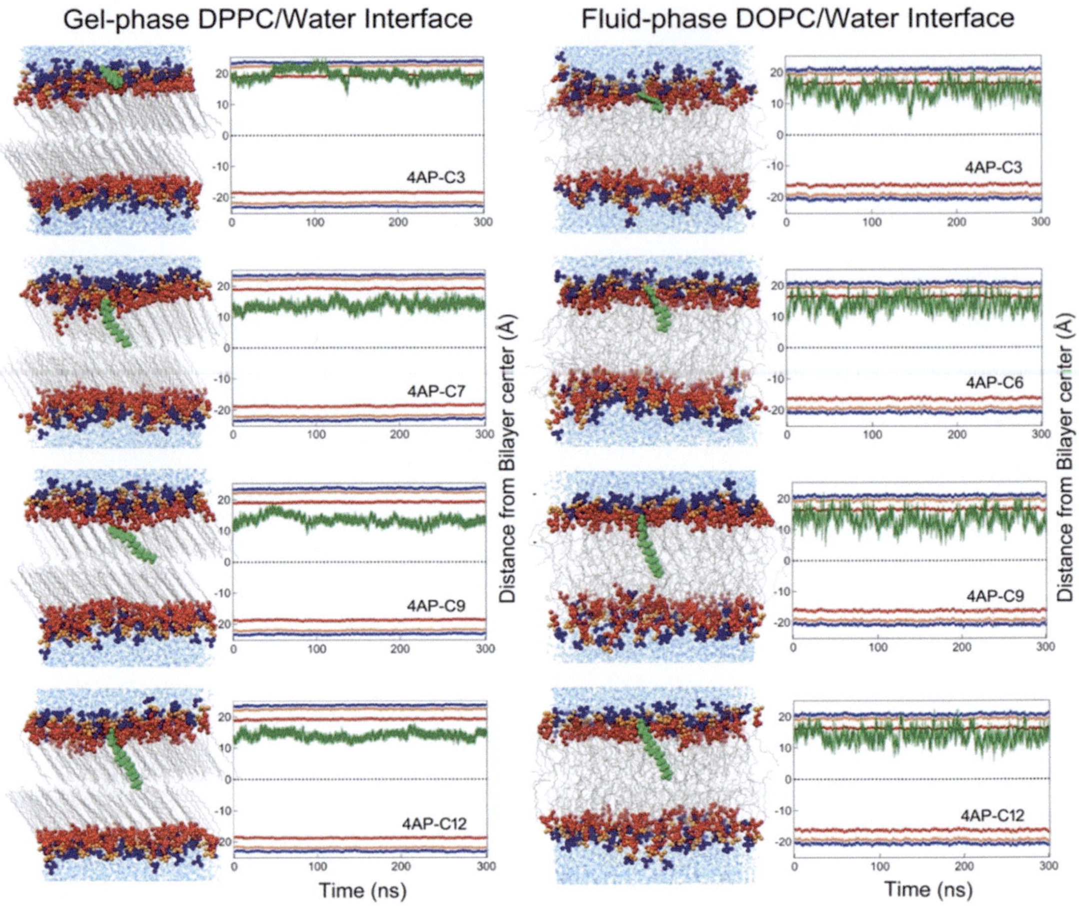

Fig. 8 Snapshots from MD trajectories, showing most likely positions of 4AP-Cn probes at the lipid/water interfaces of gel-phase DPPC (left) and fluid-phase DOPC (right) bilayers. Fluctuations in probe positions are also plotted. Color code: light blue—water molecules; Blue—choline groups, orange—phosphate groups, red—glycerol along with carbonyl groups, grey—hydrocarbon tails of lipids and green—4AP-Cn probes. Reproduced from [8] with permission from the PCCP Owner Societies

the lipid/water interfaces. Figure 8 shows the snapshots obtained from MD trajectories representing the position of 4AP-Cn probes in the lipid bilayers, and also the plots of position fluctuations of the centre of mass of 4AP moiety of probes at the lipid/water interfaces. On an average 4AP-C3 is found to stay near the carbonyl groups of bilayers, while the other probes are located much below the carbonyl region in DPPC bilayer. However, the probes show large position fluctuations at the fluid-phase DOPC/water interface.

- Position distributions of 4AP-Cn probes can be obtained for the z-axis position fluctuations of centre of mass of 4AP-Cn using "gmxtraj" command in GROMACS (Fig. 9a). Position distributions of probes in gel-phase DPPC bilayer show a systematic

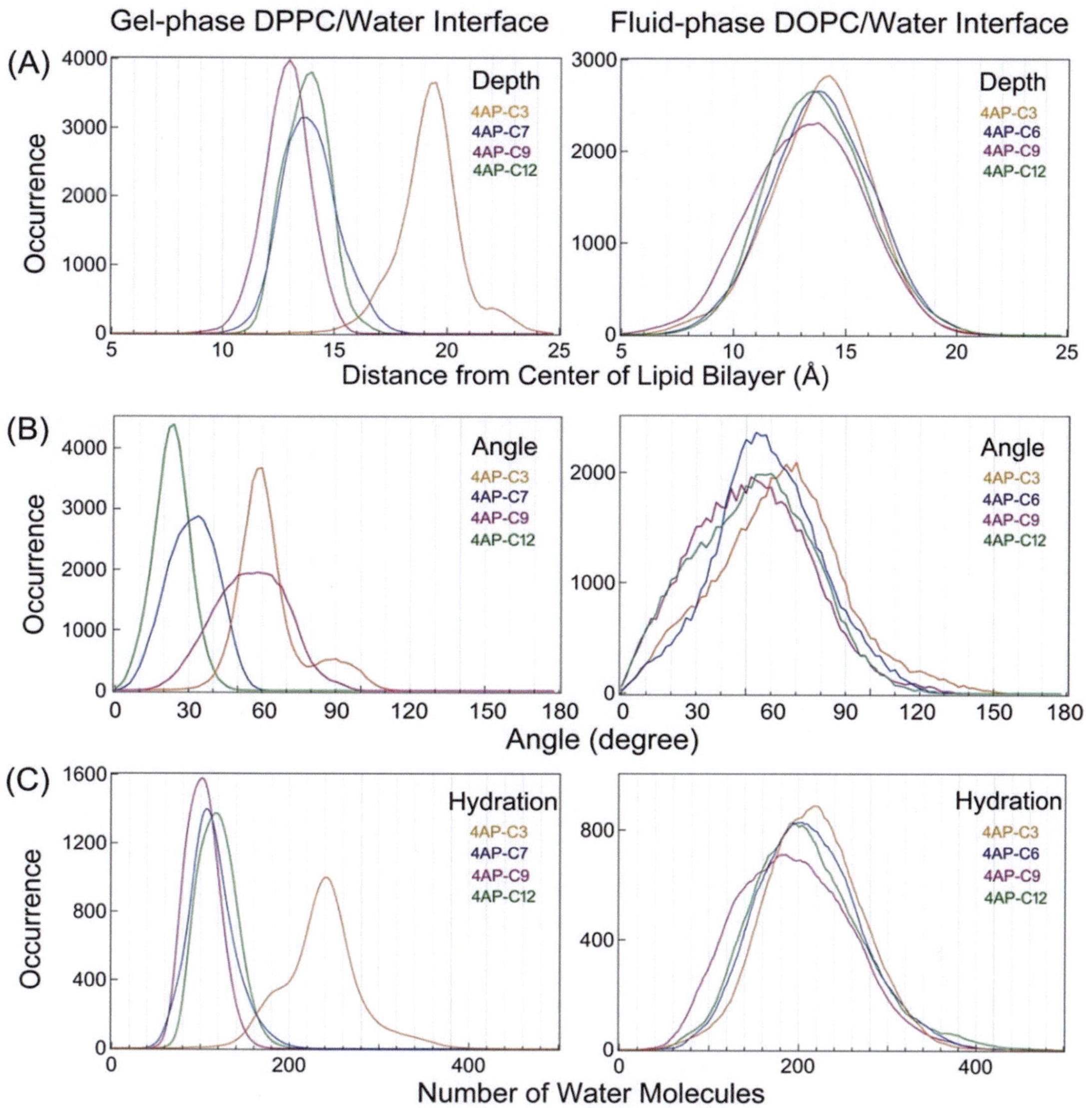

Fig. 9 Distributions of (**a**) probe position, (**b**) angle and (**c**) hydration at gel-phase DPPC/water interface (panels in left) and fluid-phase DOPC/water interface (panels in right) obtained from MD trajectories. Reproduced from [8] with permission from the PCCP Owner Societies

drift towards deeper region with increase in chain length of 4AP-Cn. This change in position shows similar trend as that of local polarity sensed by the probes at DPPC/water interface. However, probes position distributions at labile DOPC/water interface are much broader, showing only a subtle change (Fig. 9a).

- The angle distribution of the probes, relative to the bilayer normal can also be calculated from the MD trajectories by determining the angle formed between the vector connecting

N2 and C2 atoms of 4AP moiety and the normal to lipid bilayer using "gmxgangle" command of GROMACS. Figure 9b shows that at DPPC/water interface the angle distributions of probes are broader for 4AP-C3 and -C9. A second peak at ~90° with fewer occurrences is also seen for 4AP-C3 (*see* Fig. 9b). The angle distributions of other probes sharpen systematically and drifted toward smaller angles. However, the angle distributions are significantly broadened at the labile DOPC/water interface, which shows a small but systematic shift in the peak from larger to smaller values from 4AP-C3 to -C9.

- Distribution of number of water molecules within a region of 10 Å from 4AP moiety can be also calculated from MD trajectory (Fig. 9c). At DPPC/water interface 4AP-C7 has access to smaller number of water molecules (~108) than that of around 4AP-C12 (~118) (Fig. 9c). This difference in the relative hydration can arise due to the inclination angle of 4AP-C7 (~34°) which is more tilted towards lipid bilayer as compared to the inclination angle of 4AP moiety of -C12 (~24°). Further interesting properties are observed for 4AP-C3, where the effect of dual peak in the angle distribution is reflected in the (nearly) mirror-imaged distribution of water molecules at lipid/water interface of DPPC, i.e., a subset of 4AP-C3 probes with larger angle has access to less number of water molecules as compared to less inclined probes. Similarly, the 4AP-C9 probes that are located in deep regions with larger inclination angle (*ca* 58°) have less access to interfacial water molecules, resulting in less polar environment. However, the probe orientations at labile DOPC/water interface are found to be broad.

3.2.4 Relative Positions and Orientations of 4AP-Cn and Rh6G at Interface

- The MD trajectories for Rh6G in DPPC and DOPC bilayers can now be used in same way as above to obtain the *relative* position and angle distributions for the energy *acceptor* Rh6G and *donor* 4AP-Cn probes at the DPPC/water and DOPC/water interfaces.
- Figure 10 shows the relative change of positions and orientations of *donor* (4AP-Cn) and the *acceptor* (Rh6G) among which the FRET occurs. It is observed that Rh6G stays mostly in the water-rich region of the rigid gel-phase DPPC/water interface while it can penetrate into the head group region of lipids in the labile DOPC/water interface leading to broad distributions in the Rh6G positions and orientations. These simulation results can be used to calculate the closest approach of *donor–acceptor* pair at the interface and compare the same directly to the experimental results which show very good agreement [9].

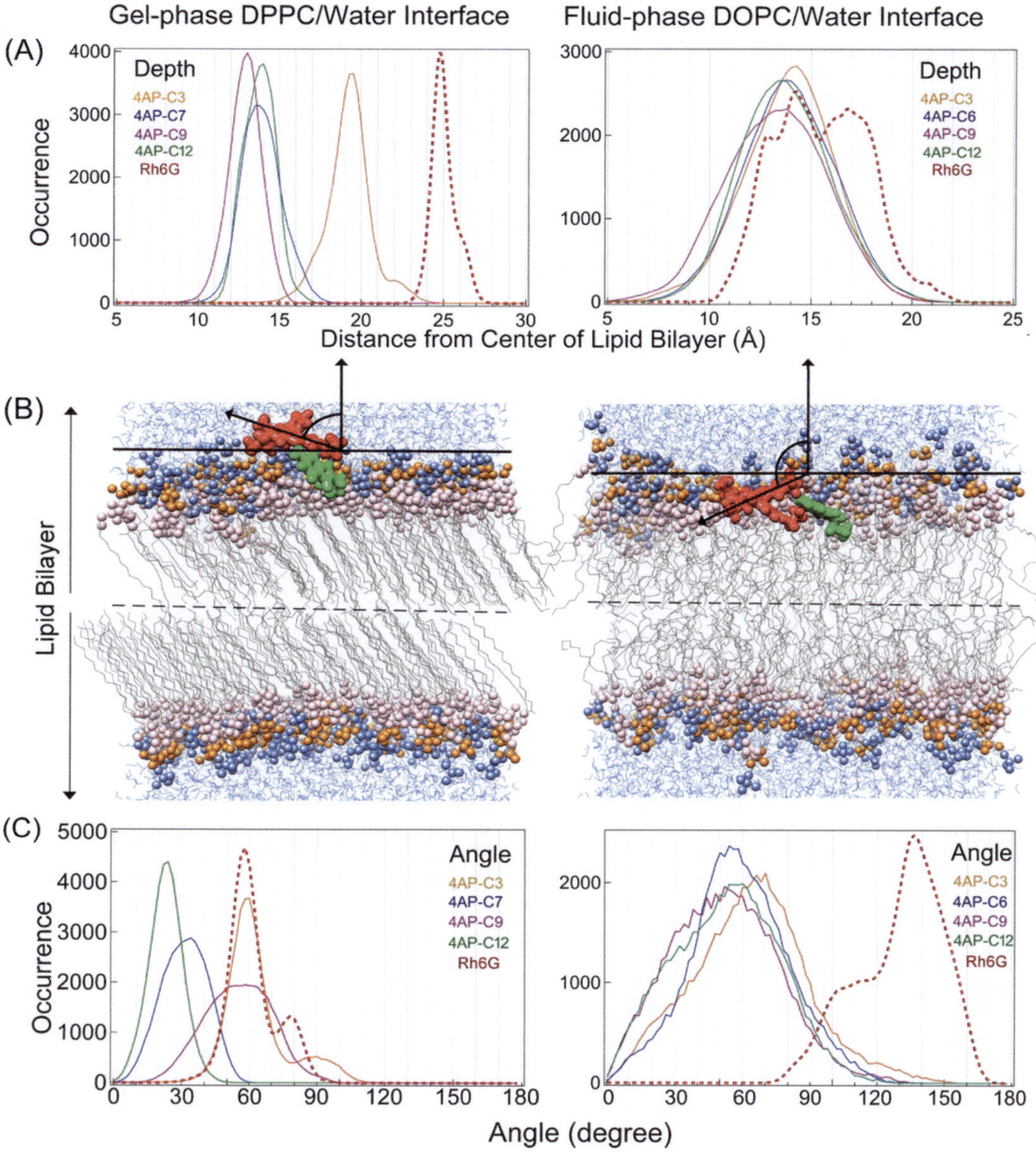

Fig. 10 (**a**) Position distributions of donor 4AP-Cn (solid lines) and acceptor Rh6G (dotted lines) at DPPC/water interface (left panel) and fluid-phase DOPC/water interface (right panel). (**b**) Cartoons showing the orientations from a MD snapshot at these interfaces. (**c**) Distributions of orientation fluctuations of donors (solid lines) and acceptor (dotted lines) at the interfaces. Reproduced from [9] with permission from the PCCP Owner Societies

3.2.5 Lipid Perturbation by Probes

- *Area per Lipid*: The extent of perturbation in the lipid bilayer by the 4AP-C probes can be observed by calculating the area per lipid, which is occupied by individual phospholipid, in the presence of the probes. To calculate this, divide the size of simulation box in X and Y direction (Box-$X \times$ Box-Y) by the total number of lipids present in a single leaflet of bilayer. These values are presented in Table 1. The values for both DOPC and DPPC

Table 1
Area per lipid in absence and presence of 4AP-Cn probes

	Area per lipid (Å²)	
Lipid bilayers	**Simulation**[a]	**Experiment**
DPPC (only)	48.8 ± 0.5	47.2 ± 0.5[b]
DPPC with 4AP-C3	49.1 ± 0.4	–
DPPC with 4AP-C7	49.2 ± 0.4	–
DPPC with 4AP-C9	49.1 ± 0.4	–
DPPC with 4AP-C12	48.9 ± 0.4	–
DOPC (only)	66.4 ± 1.2	67.4 ± 0.1[c]
DOPC with 4AP-C3	66.3 ± 1.3	–
DOPC with 4AP-C6	66.0 ± 1.2	–
DOPC with 4AP-C9	66.1 ± 1.2	–
DOPC with 4AP-C12	66.5 ± 1.2	–

Reproduced from [8] with permission from the PCCP Owner Societies
[a]Errors calculated from standard deviation of area fluctuations over entire 300 ns
[b]Data taken from [49]
[c]Data taken from [50]

bilayers in the absence and presence of the probes are found to be very similar, which are in good agreement with previous experimental results [49, 50]. This data indicate that perturbation by 4AP-Cn probes to the fluid-phase DOPC and gel-phase DPPC bilayers is insignificant.

- *Order Parameter of Lipid Chains*: Another way of checking the extent of perturbation is to calculate the carbon-deuterium order parameters (in present simulation it is carbon–hydrogen order parameter), S_{CD}, which measures the order of carbon tail of lipid bilayer [51]. This order parameter is sensitive to the chain orientation and is expressed as,

$$S_{\mathrm{CD}} = \frac{1}{2}\left\langle 3\cos^2\theta - 1\right\rangle \tag{6}$$

where, θ is the angle formed between the vector joining *i*th carbon to its hydrogen atom and the bilayer normal. Use analysis module, "g_order" of GROMACS to compute the order parameter of lipid chains. Calculate the order parameter of methylene carbons of *sn*-1 and *sn*-2 lipid tails of DOPC and DPPC bilayers with and without the probes. It can be seen that perturbation by probes in fluid-phase bilayer is almost negligible and the *sn*-1 and *sn*-2 carbon chains retain initial configuration, whereas the perturbation in rigid

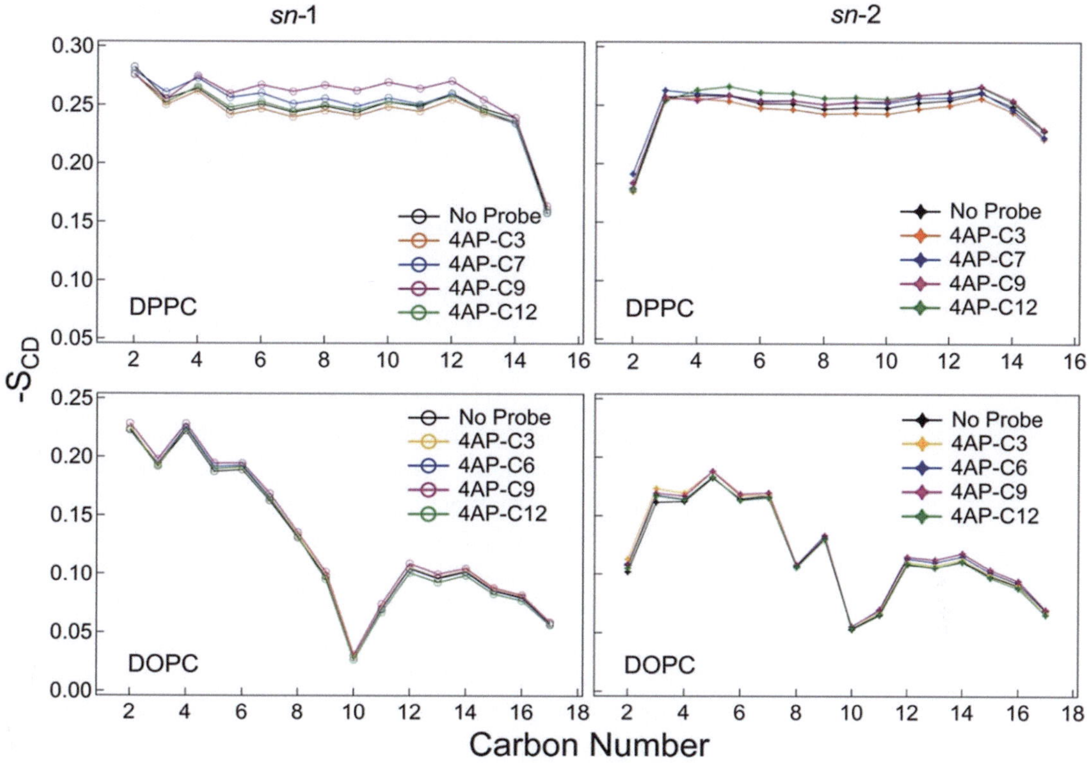

Fig. 11 S_{CD} order parameters of methylene carbons connected to *sn*-1 (panels in left) and *sn*-2 (panels in right) tail of DPPC and DOPC bilayer in the absence and presence of 4AP-C*n* probes. Reproduced from [8] with permission from the PCCP Owner Societies

gel-phase DPPC bilayer shows change of only ~3–7% (Fig. 11). These results indicate that 4AP-Cn probes barely perturb the lipid packing and are much better suited than any other ESR or fluorescent probes to explore the static and dynamic solvation properties and physiochemical characteristics of lipid membranes.

4 Notes

1. Anhydrous acetonitrile is crucial. It is treated with activated molecular sieves of pore size 3 Å (10% m/v) for 24 h followed by fresh distillation by refluxing with calcium hydride (CaH_2) for 2 h. Reaction in dry acetonitrile would occur readily but ensure that carbonate is also dry. Crush and spread the anhydrous potassium carbonate evenly in a shallow dish and dry it in an oven at 120 °C for 2 h.
2. The 4-aminophthalimide is a bit light sensitive. Therefore, it is always better to cover the flask with aluminium foil during the reaction. Check the progress of reaction through TLC using

2:1 (v/v) hexane/ethylacetate eluent mixture. To avoid any side product, always maintain the ratio of 4-aminophthalimide/alkylbromide to 1:0.9 equiv.

3. The resulting lipid film should be dry to the eye, with no remaining fluid discernible. To ensure complete drying, further dry in vacuum desiccator overnight.
4. Although TLC shows only one spot, sometimes ^{1}H NMR signal indicates presence of aliphatic impurities especially for 4AP-Cn with $n \geq 9$. To remove aliphatic impurities, wash the crystal 1–2 times with HPLC grade hexane before final use.
5. We checked the binding kinetics of 4AP-Cn with liposomes by titrating liposome with the dye, keeping 4AP-Cn concentration constant and varying the liposome concentration when preparing the vesicles. At [4AP-Cn]/[lipid] ratio 1:35, 4AP-Cn fluorescence reaches the maximum. Such choice confirms that all 4AP-Cn molecules are bound to liposome. For all steady-state measurements presented here, the concentration of 4AP-Cn was kept at 50 μM.
6. For the liposome preparation, a probe sonicator can also be used. However, water bath sonicator is better, inexpensive and does not require protective earphones. Another advantage to bath sonicator over probe sonicator is that the probe sonicator tends to heat up and cause decomposition of phospholipids at the surface of the probe and sometimes leach metal into the solution. Since the transition temperature (T_m) of DPPC is 41.3 °C, it is advised to set the temperature of bath sonicator at 60 °C.
7. Avoid plastic microcentrifuge tube for sample preparation because repeated freeze-thaw and sonication may cause solubilization of polymer tube into the lipid solution. In fact, most of glass bottles are susceptible to repeated freeze-thaw and sonication. We used 5 ml glass bottles from Borosil.
8. Note that 4AP-C2 remained unbound to DPPC bilayer at working concentration; thus, reliable data for this probe in DPPC vesicles could not be obtained.
9. Choose emission wavelengths for measuring donor decays in the blue side (10–15 nm from peak position) of donor emission, so as to avoid the rise part in donor decays which arise from slow solvation dynamics of (solvatochomic) 4AP-Cn at the interfaces.
10. Slipids (Stockholm lipid) force field can generate stable gel-phase DPPC and fluid-phase DOPC structures at room temperature. This force field was found to be the best suited compared to others for modelling gel phase of DPPC with proper area per lipid values and structures which are stable over very long simulation trajectories.

Acknowledgements

The works were supported by University Grants Commission (UGC-JNU-UPE-II, project no. 75), Department of Science and Technology (DST-PURSE) and Department of Biotechnology (DBT-BUILDER; project no. BT/PR5006/INF/153/2012). M.K.S. thanks CSIR and DBT-BUILDER, and H.S. thanks UGC for providing fellowships.

References

1. Singer SJ, Nicolson GL (1972) The fluid mosaic model of the structure of cell membranes. Science 175:720–731
2. Van Meer G, Voelker DR, Feigenson GW (2008) Membrane lipids: where they are and how they behave. Nat Rev Mol Cell Biol 9:112–124
3. Stillwell W (2013) An introduction to biological membranes. Elsevier, San Diego
4. Disalvo EA, Lairion F, Martini F, Tymczyszyn E, Frias M, Almaleck H, Gordillo GJ (2008) Structural and functional properties of hydration and confined water in membrane interfaces. Biochim Biophys Acta 1778:2655–2670
5. Haldar S, Chaudhuri A, Chattopadhyay AJ (2011) Organization and dynamics of membrane probes and proteins utilizing the red edge excitation shift. J Phys Chem B 115:5693–5706
6. Filipe HAL, Moreno MJ, Loura LMS (2011) Interaction of 7-nitrobenz-2-oxa-1,3-diazol-4-yl-labeled fatty amines with 1-palmitoyl, 2-oleoyl-sn-glycero-3-phosphocholine bilayers: a molecular dynamics study. J Phys Chem B 115:10109–10119
7. Chakraborty H, Haldar S, Chong SPL-G, Kombrabail M, Krishnamoorthy G, Chattopadhyay A (2015) Depth-dependent organization and dynamics of archaeal and eukaryotic membranes: development of membrane anisotropy gradient with natural evolution. Langmuir 31:11591–11597
8. Singh MK, Him S, Khan MF, Sen S (2016) New insight into probe-location dependent polarity and hydration at lipid/water interfaces: comparison between gel- and fluid-phases of lipid bilayers. Phys Chem Chem Phys 18:24185–24197
9. Singh MK, Khan MF, Him S, Sen S (2017) Probe-location dependent resonance energy transfer at lipid/water interfaces: comparison between the gel- and fluid-phase of lipid bilayer. Phys Chem Chem Phys 19:25870–25885
10. Marsh D (2001) Polarity and permeation profiles in lipid membranes. Proc Natl Acad Sci USA 98:7777–7782
11. Marsh D (2002) Membrane water-penetration profiles from spin labels. Eur Biophys J 31:559–562
12. Martin DR, LeBard DN, Matyushov DV (2013) Coulomb soup of bioenergetics: electron transfer in a bacterial bc 1 complex. J Phys Chem Lett 4:3602–3606
13. Kuang G, Liang L, Brown C, Wang Q, Bulone V, Tu Y (2016) Insight into the adsorption profiles of the *Saprolegniamonoica* chitin synthase MIT domain on POPA and POPC membranes by molecular dynamics simulation studies. Phys Chem Chem Phys 18:5281–5290
14. Miller AS, Falke JJ (2004) Side chains at the membrane-water interface modulate the signaling state of a transmembrane receptor. Biochemistry 43:1763–1770
15. Xiang TX, Anderson BD (2006) Liposomal drug transport: a molecular perspective from molecular dynamics simulations in lipid bilayers. Adv Drug Deliv Rev 58:1357–1378
16. Disalvo EA (2015) Membrane hydration: role of water in structure and function of biological membranes. Springer, Cham
17. Cipolla D, Shekunov B, Blanchard J, Hickey A (2014) Lipid-based carriers for pulmonary products: preclinical development and case studies in humans. Adv Drug Deliv Rev 75:53–80
18. Maruyama K (2011) Intracellular targeting delivery of liposomal drugs to solid tumors based on EPR effects. Adv Drug Deliv Rev 63:161–169
19. Forster V, Signorell RD, Roveri M, Leroux JC (2014) Liposome-supported peritoneal dialysis for detoxification of drugs and endogenous metabolites. Sci Transl Med 6:258

20. Damitz R, Chauhan A (2015) Parenteral emulsions and liposomes to treat drug overdose. Adv Drug Deliv Rev 90:12–23
21. Chattopadhyay A, London E (1987) Parallax method for direct measurement of membrane penetration depth utilizing fluorescence quenching by spin-labeled phospholipids. Biochemistry 26:39–45
22. Abrams FS, London E (1993) Extension of the parallax analysis of membrane penetration depth to the polar region of model membranes: use of fluorescence quenching by a spin-label attached to the phospholipid polar headgroup. Biochemistry 32:10826–10831
23. Menger FM, Keiper JS, Caran KL (2002) Depth-profiling with giant vesicle membranes. J Am Chem Soc 124:11842–11843
24. Kim J, Lu W, Qiu W, Wang L, Caffrey M, Zhong D (2006) Ultrafast hydration dynamics in the lipidic cubic phase: discrete water structures in nanochannels. J Phys Chem B 110:21994–22000
25. Chattopadhyay A, Mukherjee S (1999) Red edge excitation shift of a deeply embedded membrane probe: implications in water penetration in the bilayer. J Phys Chem B 103:8180–8185
26. Klymchenko AS, Mely Y, Demchenko AP, Duportail G (2004) Simultaneous probing of hydration and polarity of lipid bilayers with 3-hydroxyflavone fluorescent dyes. Biochim Biophys Acta 1665:6–19
27. Parasassi T, De Stasio G, Ravagnan G, Rusch RM, Gratton E (1991) Quantitation of lipid phases in phospholipid vesicles by the generalized polarization of Laurdan fluorescence. Biophys J 60:179–189
28. Amaro M, Filipe HAL, Prates Ramalho JP, Hof M, Loura LMS (2016) Fluorescence of nitrobenzoxadiazole (NBD)-labeled lipids in model membranes is connected not to lipid mobility but to probe location. Phys Chem Chem Phys 18:7042–7054
29. Jurkiewicz P, Olzynska A, Langner M, Hof M (2006) Headgroup hydration and mobility of DOTAP/DOPC bilayers: a fluorescence solvent relaxation study. Langmuir 22:8741–8749
30. Paul A, Samanta A (2007) Solute rotation and solvation dynamics in an alcohol-functionalized room temperature ionic liquid. J Phys Chem B 111:4724–4731
31. Ingram JA, Moog RS, Ito N, Biswas R, Maroncelli M (2003) Solute rotation and solvation dynamics in a room-temperature ionic liquid. J Phys Chem B 107:5926–5932
32. Reichardt C (1994) Solvatochromicdyes as solvent polarity indicators. Chem Rev 94:2319–2358
33. Wolber PK, Hudson BS (1979) An analytic solution to the Förster energy transfer problem in two dimensions. Biophys J 28:197–210
34. Lakowicz JR (2006) Principles of fluorescence spectroscopy, 3rd edn. Springer, New York
35. Frisch MJ et al (2009) Gaussian 09, Revision A.02. Gaussian, Wallingford
36. Cornell WD, Cieplak P, Bayly CI, Kollman PA (1993) Application of RESP charges to calculate conformational energies, hydrogen bond energies, and free energies of salvation. J Am Chem Soc 115:9620–9631
37. Wang J, Wolf RM, Caldwell JW, Kollman PA, Case DA (2004) Development and testing of a general amber force field. J Comput Chem 34:1157–1174
38. Case DA et al (2012) AMBER 12. University of California, San Francisco
39. Sousa da Silva A, Vranken W (2012) ACPYPE –AnteChamberPYthonparser interfacE. BMC Res Notes 5:367
40. Spoel DVD, Lindahl E, Hess B, Groenhof G, Mark AE, Berendsen HJC (2005) GROMACS: fast, flexible, and free. J Comput Chem 26:1701–1718
41. Jambeck JPM, Lyubartsev AP (2012) An extension and further validation of an all-atomistic force field for biological membranes. J Chem Theory Comput 8:2938–2948
42. Jorgensen WL, Chandrasekhar J, Madura JD, Impey RW, Klein ML (1983) J Chem Phys 79:926–935
43. Berendsen HJC, Postma JPM, Vangunsteren WF, Dinola A, Haak JR (1984) Molecular dynamics with coupling to an external bath. J Chem Phys 81:3684–3690
44. Hoover WG (1985) Canonical dynamics: equilibrium phase-space distributions. Phys Rev A 31:1695–1697
45. Nose S (1984) A unified formulation of the constant temperature molecular dynamics methods. J Chem Phys 81:511–519
46. Hess B, Bekker H, Berendsen HJC, Fraaije JGEM (1997) LINCS: a linear constraint solver for molecular simulations. J Comput Chem 18:1463–1472
47. Darden T, York D, Pedersen L (1993) Particle mesh Ewald: an N·log(N) method for Ewald sums in large systems. J Chem Phys 98:10089–10092
48. Parrinello M, Rahman A (1981) Polymorphic transitions in single crystals: a new molecular dynamics method. J Appl Phys 52:7182–7190

49. Tristam-Nagle S, Zhang R, Suter RM, Worthinton CR, Sun WJ, Nagle JF (1993) Measurement of chain tilt angle in fully hydrated bilayers of gel phase lecithins. Biophys J 64:1097–1109
50. Kucerka N, Nagle JF, Sachs JN, Feller SE, Pencer J, Jackson A, Katsaras J (2008) Lipid bilayer structure determined by the simultaneous analysis of neutron and X-ray scattering data. Biophys J 95:2356–2367
51. Lafleur M, Bloom M, Eikenberry EF, Gruner SM, Han Y, Cullis PR (1996) Correlation between lipid plane curvature and lipid chain order. Biophys J 70:2747–2757
52. http://www.fos.su.se/~sasha/SLipids/Downloads.html

Chapter 11

Two-Dimensional Infrared Spectroscopy of Nitrile Labels as a Tool to Probe Dynamics and Interactions in Lipid Membranes

Ilya Vinogradov and Sachin Dev Verma

Abstract

Two-dimensional infrared spectroscopy has been established as an excellent technique to study structure, dynamics, and molecular interactions in lipid membranes. Topics such as water occupation and dynamics, ion channels, electrostatic interactions, domain formation, and lipid–protein interaction in lipid bilayers have successfully been investigated. In this chapter, two-dimensional infrared spectroscopy technique, experimental methods, designing an experiment, procedures, data acquisition, data analysis, and data interpretation are discussed for nitrile vibrational labels as example probes to study dynamics and interactions in a lipid membrane.

Keywords Two-dimensional infrared spectroscopy, Vibrational spectroscopy, Nitrile label, Cyanophenylalanine, *p*-Tolunitrile

1 Introduction

Two-dimensional infrared (2D IR) spectroscopy has emerged as an excellent tool to study membrane dynamics. The most promising application of 2D IR is its use in measuring ultrafast structural and environmental fluctuations [1]. The applicability of 2D IR is rapidly growing due to the improvements in the fields of laser pulse shaping and novel detection schemes [2, 3]. Pulse shaping reduces the acquisition time remarkably, and novel detection scheme improves the signal-to-noise ratio by an order of magnitude. These new technological advancements are making 2D IR an accessible method to measure ultrafast dynamics in membranes and interaction among its constituents. 2D IR has successfully been used to understand buried water [4] and interfacial water dynamics [5], domain formation [6], and lipid–protein interactions in lipid bilayers [7].

Rajendra Prasad and Ashutosh Singh (eds.), *Analysis of Membrane Lipids*, Springer Protocols Handbooks, https://doi.org/10.1007/978-1-0716-0631-5_11, © Springer Science+Business Media, LLC, part of Springer Nature 2020

1.1 Method Prerequisites

2D IR spectroscopy is a nonlinear, ultrafast pump–probe technique that requires a sufficient understanding of optics and optical processes [8]. 2D IR measures frequency–frequency correlation functions (FFCF) of excitation (pump pulse) and detection (probe pulse) frequencies. Time and frequency are related by the Fourier transform relationship, which dictates that a short pulse results in a broad frequency bandwidth. Time resolution and frequency bandwidth are governed by the time-bandwidth product which defines the shortest pulse that can be attained for a given frequency bandwidth. An increase in time resolution results in a decrease in frequency resolution. This necessitates the use of ultrafast laser pulses in 2D IR experiments. 2D IR spectra are rich in information and are often spectrally congested. Therefore, it is required that the vibrational frequency mode of interest is adequately separated from the vibrational modes associated with the solvent and other groups in the system.

1.2 2D IR and Dynamics in Membrane

Developed primarily to study aqueous peptides and proteins, 2D IR has been established as a sensitive technique to probe the structure and dynamics of lipids and membrane proteins. Membranes, comprising thousands of lipid species and accommodating thousands of different proteins, are highly dynamic in nature. Dynamic interactions with their lipid environments lead to the function of integral membrane proteins [9–12]. For example, the timescale of the rhodopsin photocycle is modulated by the presence of cholesterol and its effect on the lateral diffusion of lipids [13]. ATPase modulation of the enzymatic activity depends on the fluidity of the surrounding lipid environment [14]. 2D IR has been employed to study electrostatic interactions [15], water occupational sites [16], association of an anchor dipeptide [17], ion channel conformational change due to phase transition [18], dynamics of high amplitude electric fields [19], and effect of cholesterol [20] on lipid membranes. 2D IR utilizes environment-sensitive probes, naturally occurring or chemically incorporated into lipids or membrane proteins, to provide information on their local structure and dynamics. Nitrile containing groups can be synthetically incorporated at different positions in lipids which can then be utilized to study dynamics at different locations in a membrane [21]. Native amino acids can be modified to have nitrile groups that can provide insights into local structure and dynamics, for example, cyano-phenylalanine.

2 Materials

1. All solvents—chloroform, any spectroscopic grade. It should be free of any nitrile containing contaminants.
2. Water—Deionized, any spectroscopic grade.

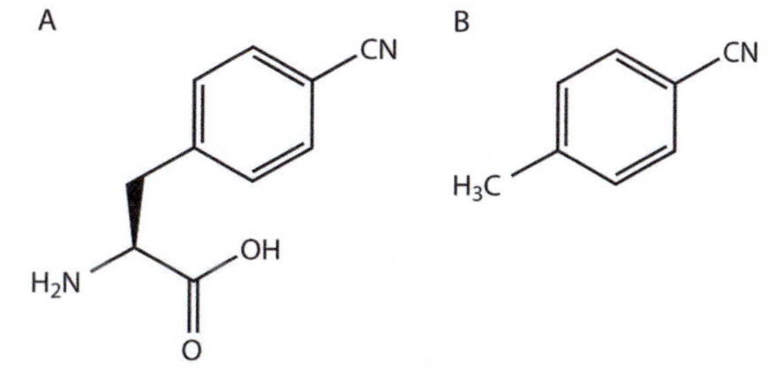

Fig. 1 Molecular Structures: A. 4-Cyanophenylalanine (Phe-CN). B. p-Tolunitrile (p-Tol-CN)

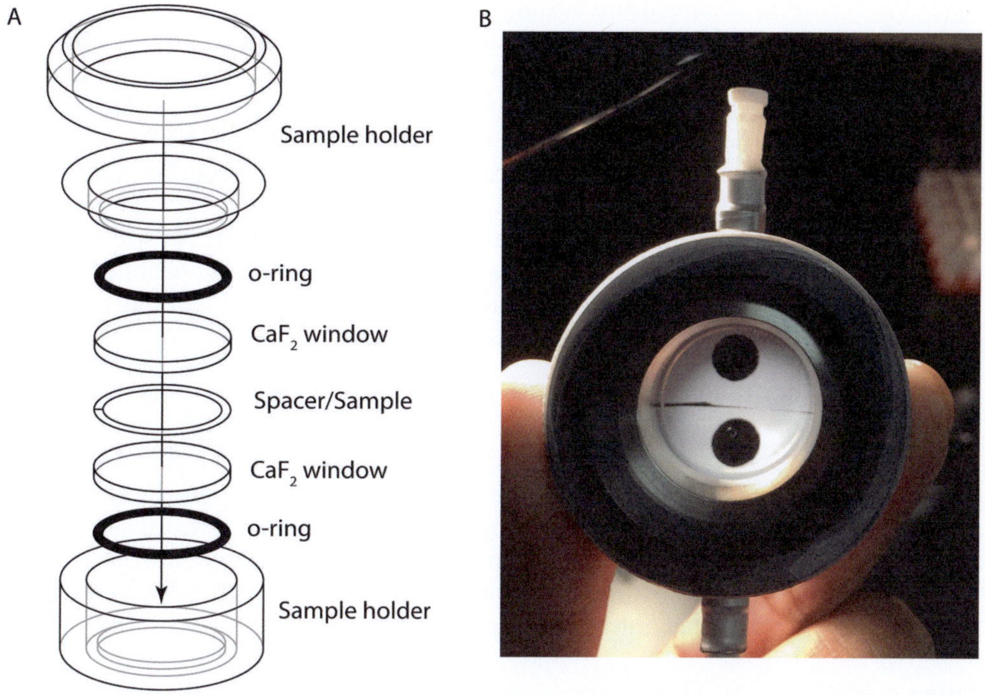

Fig. 2 A. An exploded view of a typical IR transmission sample cell. B. A photograph of an IR transmission cell designed to hold both a sample and solvent background simultaneously

3. Deuterated water— >99.8% isotopic purity for sample preparation, lower isotopic purity is okay for H-D exchange.
4. IR Labels—nitrile label in 4-Cyanophenylalanine (Phe-CN, Fig. 1a) and p-Tolunitrile (p-Tol-CN, Fig. 1b).
5. Sample Chamber—Teflon spacer sandwiched between two CaF_2 windows. To hold the windows and the sample, we use the demountable liquid sample by Harrick Scientific, CaF_2 windows of 25 mm diameter and thickness of 2 mm, and 12–100 μm thick Teflon spacers as required (Fig. 2a).

3 Methods

3.1 Sample Preparation

A stock solution of approximately 1 mL of 20 mM 4-Cyanophenylalanine (Phe-CN, 190.2 g/mol) in MilliQ purified water was prepared by weighing approximately 3.80 mg of Phe-CN in a tared vial and adding 1 mL of purified H_2O. The compound readily dissolves with swirling. The amount of H_2O added should be calculated based on the mass of Phe-CN.

Similarly, a stock solution of approximately 1 mL of 35 mM p-Tolunitrile (p-Tol-CN, 117.2 g/mol) in chloroform was prepared by weighing approximately 4.10 mg of p-Tol-CN in a tared vial and adding 1 mL of $CHCl_3$. p-Tol-CN has a low melting point and may be liquid at room temperature. In this case, we used a spatula to transfer an unknown amount of p-Tol-CN to the tared vial and used the mass difference to calculate the volume of $CHCl_3$ needed. The compound readily dissolves in $CHCl_3$.

FTIR spectra were taken with 2 cm^{-1} resolution using the same sample cell described in subsequent sections. 25 μm spacers were used for both samples. Throughout the text, we will use the term absorption and optical density (OD) interchangeably.

The FTIR requires a "background" and "sample" spectrum to calculate the absorption spectrum as:

$$\text{Abs} = -\log_{10}\left(\frac{''\text{sample}''}{''\text{background}''}\right) \tag{1}$$

The "background" spectrum was taken without the sample cell and only included the sample cell mounting (or transmission) accessory. An additional set of spectra were taken of neat chloroform and H_2O with 25 μm spacers in identical sample cells. These spectra were used as solvent backgrounds.

3.2 Basics of 2DIR

In 2D IR spectroscopy, an infrared pulse sequence is used to generate a nonlinear signal (Fig. 3) [8, 22]. A coherence is generated by the first pump pulse which then gets converted to a

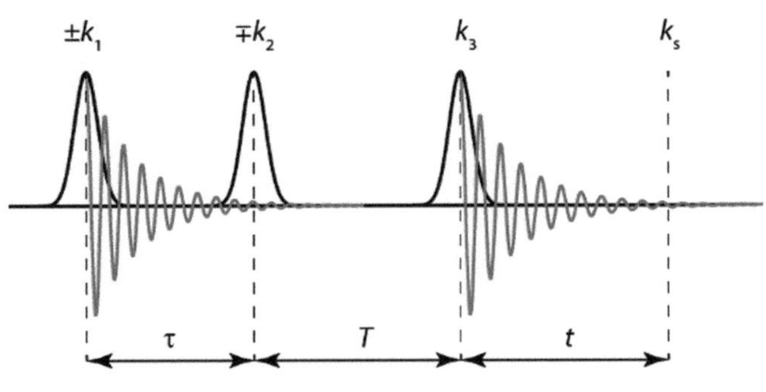

Fig. 3 Mid-IR pulse sequence used for 2D IR spectroscopy

population (or interstate coherence) by a second pump pulse. Letter *k* denotes the wavevector associated with respective pulse. The purpose of this pump pulse pair can be thought of as labelling a frequency on the ω_τ axis. Finally, after a waiting time *T*, the third pulse probes the state of the system, which generates the macroscopic polarization emitted along time *t*. This signal is typically collected by a spectrometer, which generates the frequency axis ω_t. Fourier transforming over τ generates the ω_τ axis in the 2D IR spectrum. The third pulse can be thought of as reading out the frequency of the pump-generated frequency label.

In addition, the labelled frequencies are broadened by homogeneous and inhomogeneous broadening mechanism. Homogeneous broadening is caused by dephasing induced by the oscillator's vibrational lifetime or very fast environmental fluctuations. Such dynamics effects cause the pump and probe-induced coherence to decay faster, which translate to a broader Lorentzian peak in the frequency domain. Inhomogeneous broadening is caused by environmental effects. Differences in local environments cause individual oscillators to have different instantaneous frequencies. In the FTIR, inhomogeneous broadening leads to a more Gaussian line shape. The 2D IR line shape, on the other hand, is more complex and is dynamic as a function of *T*. 2D IR can distinguish the time scales on which instantaneous frequencies change and is, therefore, capable of probing the picosecond dynamics of local environment fluctuations.

In the following sections, we will focus on a single oscillator 2D IR spectrum of a vibrational probe. These spectra represent ensemble measurements of the probe's frequencies. The statistical behaviour of the local environment's effect on the probes' vibrational frequency can be described in terms of a correlation function. This correlation function, known as the frequency–frequency correlation function (FFCF), is the main dynamic observable that we are interested in.

3.3 2DIR Setup

A typical 2DIR setup is shown in Fig. 4. The output of an ultrafast laser amplifier is fed into an optical parametric amplifier (OPA). Mid-infrared (mid-IR) laser pulses are generated with an OPA by difference frequency generation (DFG) [23]. Mid-IR pulses are then split to generate pump and probe pulses. Pump pulses are sent to a programmable Germanium (Ge)-based acousto-optic modulator (AOM) pulse shaper. An arbitrary waveform generator (AWG) sends a radio-frequency waveform to the shaper which then provides two collinear pump pulses for each input pump pulse. The delay between generated pump pulses can be tuned by the acoustic waveform in the AOM. Probe pulses go through a mechanical delay stage that can control the delay between pump and probe pulses. Pump and probe pulses, coming from different angles, are then spatially and temporally overlapped on to the sample and an

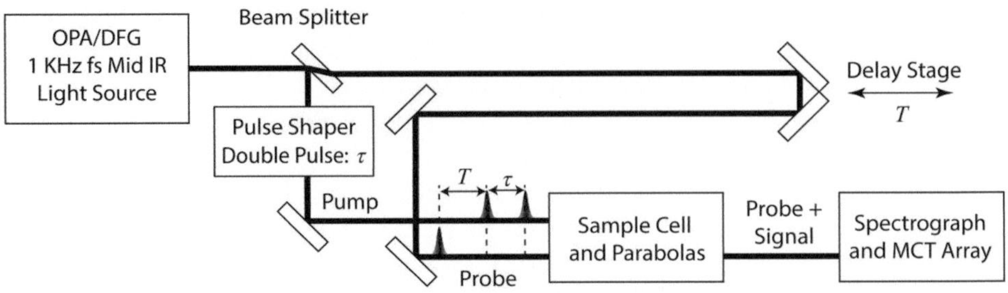

Fig. 4 A block diagram schematic of a pump–probe geometry 2D IR setup with a pulse shaper

interaction between the three laser pulses, two pump pulses and a probe pulse, takes place in the sample. Probe pulses are spectrally dispersed by a monochromator and imaged onto a liquid nitrogen-cooled mercury cadmium telluride (MCT) array detector. Spectra as a function of pump–probe delay are collected. For every pump–probe delay, second pump is delayed with respect to the first pump for a desired length of time.

3.4 Experimental Procedure

When designing a system for 2D IR spectroscopy, several considerations need to be made. Often, the biological system needs to be partially tailored to the 2D experiment. Design considerations of such a system include: (1) the choice of the vibrational probe, (2) the target concentration of the vibrational probe, (3) the solvent, (4) the sample cell and (5) the experimental parameter that is being varied. Below, we will consider each of these design points in detail with a focus on the nitrile (CN) label.

3.4.1 Choice of the Probe

The choice of the vibrational probe needs to be done in conjunction with the other system design points. Considerations for vibrational probe choice include the following. First, the probe's oscillator strength should be sufficiently strong, i.e. how much signal it will generate for a given probe concentration. Second, the probe's vibrational transition frequency should be in a sufficiently clean spectral region, i.e. whether the lipid, solvent, and protein (if present) vibrational modes generate a strong background in the 1D IR spectrum. Third, it must be reasonably possible to synthetically incorporate the label into the system. Finally, the vibrational label must be sensitive to the experimental parameters being varied. All the above criteria must be met in order to generate interpretable 2D IR data. Several reviews are available that provide more detail on the above topics [24–27].

3.4.2 Target Concentration of the Probe

The target concentration of the vibrational probe needs to be high enough in order to be able to acquire the desired 2D IR signal in a reasonable time frame. The waiting-time-dependent 2D IR experiment with the pulse sequence described above is generally time-consuming. In order to be able to extract information about the 2D IR line shape, both the T and τ delays (as shown in Fig. 3) must be scanned while the third delay, t, is usually acquired directly with a spectrograph. 2D IR signals, being induced by nonlinear optical interactions, are usually weak and require substantial averaging. Furthermore, label concentrations often need to be on the order of a few millimolar (mM) or lower for the sample conditions to be biologically relevant. The end result is that publication-quality 2D IR spectra typically take between several hours to days to acquire for a particular condition. A general rule of thumb is that if the peak can be clearly distinguished in the FTIR, a usable 2D IR spectrum can be acquired with enough averaging.

The hard limit on the vibrational probe concentration is the signal strength relative to the background. The background usually comes from 2D signals from lipids or peptide, or other nonlinear processes, i.e. coherence artifacts [28]. If a strong background is present in the 2D IR spectra, the data processing methods described in the following sections will not be reliable. Fortunately, the 2D IR signal strength, being a third-order nonlinear signal, scales as the fourth power of the transition dipole moment [29]. The linear 1D IR signal, on the other hand, scales as transition dipole moment squared. This means that the 2D IR signal, as compared to the 1D IR signal, will scale quadratically with oscillator strength. Solvent or lipid modes that contribute to backgrounds in the 1D IR spectrum usually absorb strongly because of their high concentration, but otherwise have weak oscillator strengths. Given that both 1D and 2D IR signals scale linearly with concentration, solvent or lipid backgrounds are usually weaker and are less of an issue in 2D IR spectra as compared to 1D IR spectra. Therefore, labels with stronger transition dipole moments are preferable.

As an example, consider the -CN label in cyano-phenylalanine (Phe-CN). Figure 5a–c shows the FTIR and 2D IR spectrum of 20 mM Phe-CN in neat H_2O. The FITR spectrum has a prominent background of OD 0.5, which is ~80× larger than the absorption of the label itself (~6 mOD). The 2D IR spectrum, on the other hand, is nearly background free. Similarly, Fig. 5d–f shows FTIR and 2D IR spectra for the –CN label in p-Tolunitrile (pTol-CN). Note that at very low concentrations, the H_2O background will begin to dominate and distort the 2D IR spectrum. One potential workaround for this problem is subtract out a background 2D IR spectrum that contains the same biosystem components but without the vibrational label. Such a method comes at a substantial cost. First, the data acquisition time for any particular condition doubles. Second, the act of subtracting noisy foreground data from noisy background data decreases the SNR by a factor of $\sqrt{2}$. Third, the presence of systematic error in sample preparation and 2D IR setup alignment often requires that the background and foreground data

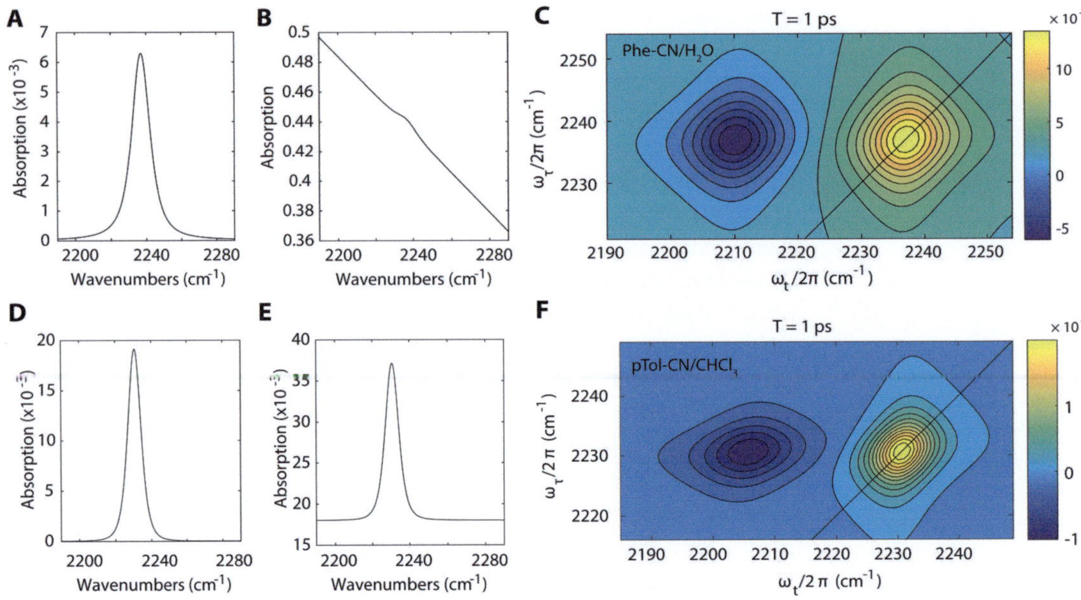

Fig. 5 Simulated spectra of the nitrile label. (**a**) FTIR of 20 mM Phe-CN in neat H_2O with background removed. (**b**) 20 mM Phe-CN including H_2O background absorption. (**c**) 2D IR spectrum at $T = 1$ ps for 20 mM Phe-CN in H_2O. Notice that the H_2O background is greatly reduced (see color bar for scale). (**d**) FTIR of 35 mM p-Tol-CN in neat $CHCl_3$ with background removed. (**e**) 35 mM p-Tol-CN including $CHCl_3$ background absorption and reflective loss from the CaF_2 windows. (**f**) 2D IR spectrum at $T = 1$ ps for 35 mM Phe-CN in $CHCl_3$. All spectra are with a 25 μm spacer

be acquired in the same data acquisition time period and the same sample cell (an example of such a sample cell is shown in Fig. 2b) and with an automated sample cell stage to switch between the two samples.

3.4.3 Solvent

The third important consideration is the sample solvent. The main consideration in the solvent choice is that the solvent should not be strongly absorbing in the IR region of interest. A strongly absorbing solvent will deplete the pump, thereby lowering the signal strength; deplete the probe, thereby increasing the relative contribution of detector noise to the SNR; and can act as a strong background in the 2D IR spectrum. For biosystems, the two most common solvent choices are water (H_2O) and heavy water (D_2O). Note that it is also possible to study model lipid systems with organic solvents, especially if one is interested in reverse micelles [30]. We will only consider the advantages and disadvantages of H_2O and D_2O.

Figure 6 shows the spectra of neat H_2O and D_2O in a 25 μm thick sample cell. Both solvents are strongly absorbing across the full mid-infrared spectrum. H_2O is useful in the spectral region around 3.6–5.5 μm (1800–2800 cm^{-1}). H_2O is ideal for many vibrational labels, including nitriles and metal carbonyls, because

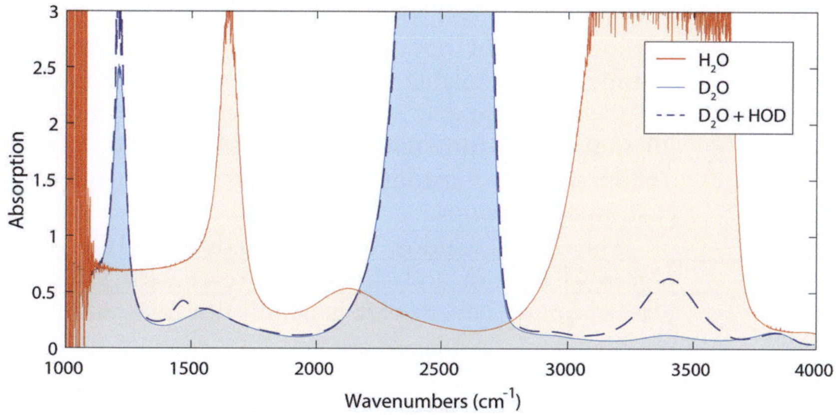

Fig. 6 FTIR spectra of H_2O (red), D_2O (solid blue), and D_2O with HOD contamination (dashed blue) with a 25 μm spacer. The noise below 1100 cm^{-1} is due to CaF_2 absorption

this region is clear of most strongly absorbing lipid modes and protein modes. In addition, and in our experience, 4–5 μm mid-IR femtosecond (fs) pulses are particularly efficiently generated by DFG of the signal/idler output of an OPA using an $AgGaS_2$ crystal. The disadvantage of using H_2O is that it strongly absorbs near vibrational modes that are intrinsic to biomolecules. Such modes include the phospholipid ester and phosphate modes [31, 32], and peptide amide modes [33].

To overcome this disadvantage, we use D_2O as the solvent, which is useful in the spectral region between 2.5–3.5 μm (4000–2850 cm^{-1}) and 4.5–7.4 μm (2200–1300 cm^{-1}). D_2O may also be useful for studying low-frequency phosphate modes, around 10 μm, although 2D spectroscopy is generally challenging (but still possible) at such long wavelengths due to issues with optics and difficulties in generating sufficiently high powered mid-IR light at longer wavelengths. The use of D_2O also introduces several other difficulties. First, one must carefully exchange all free protons in the sample to avoid the formation of HOD, which has a prominent absorption around 6.8 μm (1475 cm^{-1}) (see dashed line in Fig. 6). In addition, the prepared sample must be kept in low-humidity environments to avoid H-D exchange with air moisture. We generally use 99% isotopic purity D_2O for sample deuterium exchange and >99.8% purity D_2O ampules for actual sample preparation. Actual sample cells are stored, refrigerated, in desiccators or jars filled with desiccant, which minimizes air moisture H-D exchange. Such samples usually last about one week before H-D exchange becomes a significant problem. Note that we have successfully frozen exchanged and lyophilized samples for several years. Exchange can be readily monitored by FTIR (*see notes for deuterium exchange*).

3.4.4 Sample Cell

For the sample cell, two main considerations are the sample cell thickness and the optical window material. Many sample cells are commercially available. To hold the windows and the sample, we use the demountable liquid sample cell by Harrick Scientific, dia. 25 mm and 2 mm thick CaF_2 windows, and 12–100 μm thick Teflon spacers. Custom sample cells are also often designed for custom applications.

The optical window material needs to be IR transparent in the region of interest and should not react with the sample. Plastic, glass or quartz cuvettes are generally not transparent in the mid-IR region where most labels are available. A popular choice of window material is CaF_2, which is IR transparent until ~ 10 μm (10,000 cm^{-1}). CaF_2 is a relatively soft and brittle material that scratches and breaks easily. For 2D IR measurements, care should be taken to avoid scratches because they will act as a strong source of pump–probe scatter, which will generate a prominent interference pattern on top of the 2D IR signal. Pump–probe scatter will be described in more detail later.

The allowable sample thickness is limited by the sample optical density and by the pump–probe overlap volume. The overlap thickness limitation is determined by the volume in which the pump and probe beams are physically overlapped. Unless the 2D IR setup is specially designed for a low pump–probe incidence angle or co-propagating pump and probe beams, the thickness limitation will be on the order of several hundred micrometers (this depends on the pump–probe incidence angle, which is 15° for our setup). Samples thicker than this limit will not generate additional signal. The optical density (or amount of IR absorption) of the sample usually poses a much more severe limitation, unless using very low-absorbing solvents such as chloroform. In H_2O and D_2O, the most practical sample thickness is around 25–50 μm. Sample beyond the initial 50 μm will not significantly contribute to signal because of pump depletion (the intensity falls off exponentially). For example, the solvent background for Phe-CN in water has a background absorption of 0.5 for a 25 μm spacer, which will absorb 68% of the pump. Doubling the spacer thickness to 50 μm will absorb 90% of the pump. Other regions of the H_2O or D_2O spectral region are more strongly absorbing.

If the sample is viscous or difficult to work with, it may be desirable to use a thicker spacer, e.g. 50 or 100 μm. In this case, the sample may absorb >90% of the incoming probe. One must make sure that the detector (usually an MCT detector) detects enough of the leftover probe light so that the SNR is not detector noise limited (unless using referencing schemes to increase SNR [3]). To maximize the measurement SNR, the probe light intensity should nearly saturate the MCT detector pixels. Note that these detectors can be highly nonlinear, especially if used with low gain settings [3]. In this case, the MCT detector should be used near the top of the linear range, unless the nonlinearity is corrected for.

In general, FTIR and 2D IR sample cells for biomolecules require small volumes on the order of 50 μL or less. For 2D IR spectroscopy, the sample volume requirement can be even smaller. A general rule of thumb is that for a dia. 25 mm spacer with 2 mm of Teflon on the edge, the sample volume should be the same as the thickness. For example, such a 25 m thick spacer would require 25 μL of the sample. In reality, the 25 μm spacer requires a minimum sample volume of 8.7 μL. In practice, the extra sample volume is required to help remove bubbles. Finally, for more customized applications, one can cut out a spacer from a thin Teflon sheet. As an example, the sample cell shown in Fig. 2b used <1 μL of the sample in each sample compartment (*see notes for sample loading*).

3.4.5 Parameter Under Investigation

The fifth and final important consideration for 2D IR spectroscopy is the parameter that is being varied. The reason we include this consideration is because 2D IR spectra for a single sample condition seldom lead to a clear conclusion (yet alone a publishable study). There are of course notable exceptions [34], even in recent years [35]. 2D IR spectra are fully capable of providing large amounts of quantitative data. Unfortunately, connecting that quantitative data to non-spectroscopic experimental parameters usually requires a significant computational undertaking. Therefore, it is much easier to use the strategy of changing an experimental parameter in order to correlate that parameter with the observables extracted from the 2D IR spectra. Usually, this is done qualitatively, but more quantitative interpretations are also possible given a good model.

The spectroscopic observables that are directly extractable from a single oscillator spectrum include the transition dipole moment, transition frequency, vibrational lifetime (or T_1 time), the pure vibrational dephasing time (or T_2^* time), and frequency–frequency correlation function (FFCF) [8]. As described earlier, the FFCF is one of the dynamic observable of interest in 2D IR. We will go into more detail into how to extract the FFCF from 2D IR spectra. A non-exhaustive list of experimental parameters that the above observables can be correlated with include: temperature, pH, ionic strength, lipid composition, lipid size (e.g. vesicle size) [36], membrane fluidity, and label position. For qualitative comparison, these parameters are also varied for a model system. The model system usually consists of a closely related single-molecule label (e.g. p-Tol-CN as a mimic for Phe-CN) in a set of solvent conditions that are believed to mimic the local environment of the more complex system. If the model system attains results that are qualitatively close to those of the more complex system, one is usually able to infer that the local environments are similar. Using model systems in this way is advantageous because model systems are usually more computationally tractable and may provide a clearer physical explanation for the observed results.

3.5 Data Acquisition

Before explaining the data analysis for 2D IR spectra, we will mention some considerations related to the actual data acquisition process. First, we will give a procedure for spatially and temporally overlapping the pump and probe beams on the sample. Second, we will consider pump–probe scatter. Third, we will mention the rotating frame as implemented with a pulse shaper. Finally, we will provide some general guidelines for setting data acquisition parameters.

3.5.1 Pump–Probe Overlap

To spatially and temporally overlap the pump and probe beams, we recommend the following procedure.

Step 1: Aligning an infrared beamline is typically done with a HeNe tracer laser. A pair of irises and a set of corresponding mirrors should be used to ensure that the tracer laser is collinear with the infrared beam before entering the 2D IR setup. Both the pump and probe beams should be roughly collinear to their respective HeNe beamlines. Irises and mirrors should be added as necessary to ensure this.

Step 2: Once both the pump and probe beam paths are well defined and collinear with the HeNe tracer, place a pinhole (dia. ~100 μm, depending on the beam focus) at the sample position. Move the pinhole in the X (parallel to the table), Y (perpendicular to the table), and Z (direction of probe beam propagation) directions until the probe HeNe beam throughput is maximized.

Step 3: Switch to the infrared beam and use the MCT detector to move the pinhole (X, Y, and Z) until the actual probe beam intensity is maximized. If the correct focus position is known, then the pinhole Z-position should be kept at this focus position (in case the probe collimation is incorrect). Once the pinhole position is set, it should not be moved until this procedure is complete.

Step 4: Switch back to the HeNe and use a steering mirror for the pump beam to align the pump HeNe through the pinhole. Do not move the pinhole.

Step 5: With the pump HeNe going through the pinhole, place a detector that can detect the infrared pump beam (such as a pyroelectric detector) directly behind the sample.

Step 6: Switch to the infrared pump and use the steering mirror to maximize its throughput, as monitored by the pump detector in the previous step.

Step 7: The pump and probe beams should now be spatially overlapped. To find the temporal overlap (i.e. time zero or T_0) do as follows:

Step 8: Place a Ge semiconductor plate (usually used as an IR neutral density filter) at the sample position.

Step 9: Monitor the pump–probe signal while changing the delay. If the probe arrives after the pump, a strong signal will be present. If the probe arrives before the pump, no signal will be present. The signal that is generated by the semiconductor has a very long lifetime compared to any dynamics probed in biological systems.

Step 10: Set the delay to the onset of the pump–probe signal. This is a rough estimate of T_0.

Step 11: Fine-tuning of the T_0 can be done using a dielectric material (such as a CaF_2 window) and scanning for the coherence artifact. Alternatively, use a scattering source (such as a pinhole) and scanning the pump–probe delay until the pump–probe scatter interference has an infinite period.

For a pump–probe geometry setup, knowing the T_0 to within the <½ of the pulse duration is sufficient (e.g. ~100 fs for a 200 fs pulse). In our experience, the above method can find the T_0 to <10 fs precision. For BOX-CARS photon echo setups, the T_0 needs to be known with extreme precision (to a fraction of an optical cycle) as this will determine the phase of the acquired signal.

3.5.2 Pump–Probe Scatter

Pump–probe scatter is generated when a portion of the pump is scattered into the probe direction. Because the two beams are coherent and maintain a (nearly) constant phase relationship with each other, they will interfere. The interference pattern is dependent on both the frequencies and delays of the pump–probe beams as $\cos\omega T$, where ω is the angular frequency and T is the pump–probe delay. For 2D IR spectroscopy, two pump pulses generate additional scatter which is dependent on the pump-pulse delay τ. After Fourier transforming over τ, the double-pump–probe scatter will appear as a strong artifact along the diagonal of the 2D IR spectrum. The diagonal spectrum of the scatter will also change with the pump–probe delay T.

Fortunately, the pulse phase dependence of the pump–probe scatter and 2D IR signal is different. One can flip the phase of the 2D IR signal by changing the relative phase between the two pump pulses. The phase of the pump–probe scatter can be flipped by changing the absolute phase of the pump–pulse pair with respect to the probe pulse. Using a pulse shaper, the absolute and relative phases of the pump pulses can be controlled for every laser shot. For setups without a pulse shaper, phase cycling can be well approximated by a wobbling transmissive optic [37]. Implementing such a phase cycling scheme will allow one to significantly suppress pump–probe scatter and additional artifacts in the 2D IR spectrum [38].

The ability to suppress pump–probe scatter is largely limited to the phase stability between the pump and probe beams. For mid-IR, pump and probe beams derived from the same mid-IR beamline are usually phase stable because the wavelength is relatively long compared to shot-to-shot beam path length differences that are possible. The exception is if the laser pointing-stability is very poor, then large shot-to-shot path length differences may be introduced. With mid-IR AOM-based pulse shapers, the pump–probe phase stability is largely limited by the phase stability of the acoustic waveform [39]. Therefore, pump–probe scatter can be suppressed to a large extent, but if possible, one should minimize it at the sample.

3.5.3 Rotating Frames

With the use of pulse shaping, one can also implement a rotating frame [38, 40]. This method shifts the ω_τ zero frequency to a new rotating frame frequency and reduces the number of τ delays one must acquire. The rotating frame is implemented by introducing an appropriate τ-dependent relative phase between the pump pulses. The effect in the time domain is that the 2D IR signal no longer oscillates at the infrared frequency of ω_t. Instead, the oscillation will be at the difference frequency between the rotating frame and the ω_t being observed. This allows for a larger τ time spacing to sample the decay of the signal, thereby requiring fewer points. After Fourier transforming, the ω_τ frequency will be relative to the rotating frame frequency. In the extreme case, the rotating frame can be set to the observed ω_t. In this case, no oscillations will be present and the signal's decay can be measured directly. Note that usually several ω_t are observed simultaneously and normally it is only possible to set one rotating frame, so only one frequency will truly be free of oscillations. Analogous methods to the rotating frame exist when a pulse shaper is not available called under-sampling [41] and the quasi-rotating frame [42]. The artifact that can arise from the rotating frame is discussed in notes section.

3.5.4 Data Acquisition Parameters

Finally, we will provide a few general guidelines for data acquisition parameters. Data acquisition parameters include the spectrometer parameters, time delays (τ and T) and the choice in the rotating frame (discussed above and in the notes).

The two main spectrometer parameters of interest are the spectrometer resolution and spectral scan range. The reason these are worth mentioning is because mid-IR detectors do not have the same high pixel count that most visible detectors have. Therefore, this parameter choice is usually a balance between spectral resolution and range, which is usually changed by switching between gratings with different groove densities. For the line-shape analysis described in the next section, the spectrometer resolution needs to be good enough so that at least 2–3 pixels are present above the half maximum of the peak. In addition, we recommend that the spectral range be wide enough to encompass the full vibrational mode and its anharmonic peak in a single grating position. This will save considerably on data acquisition time by not having to scan multiple grating positions.

The time delay τ should be equally spaced and be long enough to include the full dephasing 2D IR signal. The longest τ delay can be estimated by using a model compound in a neat solvent if the sample signal is too weak to detect without substantial averaging. Generally, labels in biosystems will exhibit somewhat faster dephasing because of inhomogeneous broadening. One may also use a window function (also known as apodization) to collect up to a shorter τ delay, but at the cost of resolution. The time spacing between τ points is determined by the maximum frequency difference between ω_t and the rotating frame. The spacing should be at maximum one-half of the period that corresponds to this frequency difference (this is the Nyquist limit). For example, if the frequency difference is 200 cm^{-1}, the maximum point spacing will be:

$$\frac{1}{2} \times \frac{1}{200\ \text{cm}^{-1}} \times \frac{1\ \text{s}}{3.00 \times 10^{8}\ \text{m}} \times \frac{1\ \text{m}}{100\ \text{cm}} \times \frac{10^{15}\text{fs}}{1\ \text{s}} = 83.5\,\text{fs} \quad (2)$$

We recommend four or more points per period, which will give around 40 fs spacing. We typically use 30 fs.

The choice in waiting time T delays is less strict compared to τ. There is no good rule of thumb for the choice of maximum T because it is highly system dependent. The limits on the maximum value of T are largely determined by whether one can acquire signal at that delay. Through the projection-slice theorem, the integrated 2D signal along ω_τ is the same as the pump–probe spectrum at the given delay T. One may use the vibrational lifetime or the pump–probe spectrum (which will be a function of T and ω_t) to estimate the longest possible T needed for the FFCF analysis function described in the next section. The spacing between T points will depend on your pump and probe pulse duration. Dynamic features faster than the pulse duration will not be resolvable. We usually use a minimum step of 200 fs for a 150 fs (FWHM) pump pulse. It is also common practice to use a close-to-logarithmic scale. Usually, dynamics in the first ps require finer point spacing while dynamics after 1 ps can use a coarser point spacing.

3.6 Data Analysis

After data have been collected, a popular line shape analysis method which extracts the FFCF is the center line slope (CLS) analysis [43, 44]. The CLS analysis consists of fitting a line through the local maxima (in our sign convention) of spectral slices near the peak. The spectral slices are parallel to either the ω_t or ω_τ axis. With several approximations, the slope (or inverse slope) of the line versus T represents the normalized FFCF. Below we will give the procedure on how to use CLS analysis. This procedure will focus more on the technical aspects of CLS. For theory and further detail

Table 1
CLS analysis methods for different 2D IR plotting conventions

2D data *x*-axis	Slices axis	Analysis method	Slope as plotted	(*x*, *y*) for linear fit
Probe (ω_t)	Probe (ω_t)	CLS	Inverse	(ω_τ, ω_t)
Probe (ω_t)	Pump (ω_τ)	Inverse CLS	Normal	(ω_t, ω_τ)
Pump (ω_τ)	Probe (ω_t)	CLS	Normal	(ω_τ, ω_t)
Pump (ω_τ)	Pump (ω_τ)	Inverse CLS	Inverse	(ω_t, ω_τ)

into the procedure, we recommend the original papers [43, 44] and a comparison against other line shape analysis methods [45].

There are two variants of the CLS analysis. One variant, known in the original publication as inverse CLS, takes slices that are parallel to the ω_τ axis [43]. The second variant takes slices parallel to the ω_t axis and is known in the original publication as just CLS [44]. Whether one takes the inverse of the slope or just the slope depends on the convention used for plotting the 2D data. The procedure for CLS and inverse CLS is as follows:

Step 1: After Fourier transforming the data, select a box around the peak of interest. The box should include at least three data points along the slice direction. Several data points should be included in the other direction. The exact bounds of the box will require some iterative tweaking, as will be described below. A good rule of thumb is to avoid the anharmonic peak (unless one is performing CLS on the anharmonic peak) and try to avoid the tails of the peak.

Step 2: Take the subspectra within this box. Subdivide each subspectrum into spectral slices that are parallel to the desired axis.

Step 3: Fit the resulting 1D spectra to a functional form that will interpolate the maximum peak position for that slice. Popular fitting functions include a Gaussian and a cubic polynomial. We usually choose the three most maximum points which exactly define a cubic polynomial and extract the maximum ω_t or ω_τ values. We find that this three point gives the most robust estimate of the maximum. Note that this three-point method is exactly equivalent to a cubic spline interpolation. In practice, however, one may need to try several fitting methods to get a good set of ω_t and ω_τ coordinates.

Step 4: After extracting the maxima, fit the ω_t and ω_τ coordinates to a line. The choice of which frequency axis will be the *x*-axis for the fit depends on the convention used for plotting the spectra and whether one is performing inverse CLS analysis or just CLS analysis. An intuitive way to make the correct choice is to choose the axis combination that gives decreasing slope with longer *T*. Alternatively, one can use the guidelines summarized in Table 1.

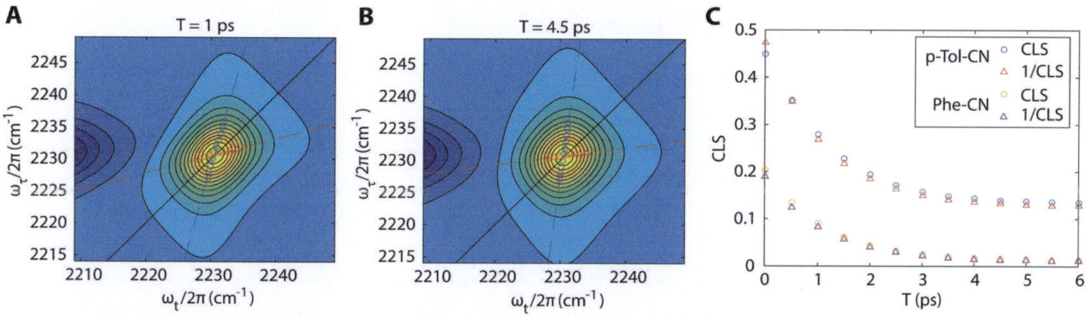

Fig. 7 CLS and inverse CLS analysis. (**a**) 2D IR spectrum showing the CLS (blue) and inverse CLS (red) points (in circles) and fit (dashed line) at 1 ps for p-Tol-CN in $CHCl_3$. (**b**) The same analysis as in A except for $T = 4.5$ ps. (**c**) CLS and inverse CLS values for different waiting times T for 35 mM p-Tol-CN in $CHCl_3$ and 20 mM Phe-CN in H_2O

Step 5: Evaluate whether the steps above generated a reasonable CLS (or inverse CLS) curve. If the curve does not look reasonable, choose a different bounding box and repeat the above procedure from **Step 1**. Unfortunately, the choice in bounding box and evaluation of whether the CLS curve is reasonable are somewhat arbitrary. One good metric for such an evaluation is to look at the linear fits. The set of local maxima points for each spectrum should generate a curve that is nearly linear. It should not contain any obvious outliers and should have any obvious shape minimized, such as "s"- or "c"- shapes. We usually try several different bounding boxes and local maxima interpolation methods. We finally choose the box/interpolation method that minimizes the linear fit residuals for all T times.

The CLS analysis provides an estimate of the normalized FFCF. It is also possible to estimate the magnitude of frequency fluctuations to extract the unnormalized FFCF. The procedure is more complex and is outlined in the original paper [43].

We recommend using CLS analysis instead of inverse CLS analysis. We find that normal CLS gives more repeatable results and is less susceptible to noise. This is likely related to the spectrometer's pixel correlation. The major noise source in pump–probe geometry 2D IR spectroscopy comes from laser shot-to-shot intensity fluctuations. Since a pixel array reads out the probe spectrum in a single shot, the apparent pixel-to-pixel noise (along ω_t) will be usually lower compared to the noise along τ, whose delays are collected for different laser shots. After Fourier transforming, excessive shot-to-shot noise and the effects of pixel correlation will appear as a horizontal (or vertical) banding noise if ω_t (or ω_τ) is plotted on the x-axis. Assuming the pixel correlation does not introduce any spectral distortions, slices parallel to ω_t may provide a better estimate of the peak local maxima when the peak is composed of a few data points.

An example of the procedure above is shown in Fig. 7. The simulated data give an exact fit for the CLS analysis but do not show the importance of the finer points described in finer points described above. An alternative approach, to circumvent the arbitrary choice in the bounding box, is to perform a full response function fit. This involves developing a quantum mechanical model and simulating the 2D IR spectrum. Key model parameters are varied until a good fit is achieved. The simulated spectra of a single oscillator can be in excellent agreement with experimental data, but the model parameters are usually not robust due to overparameterization. In addition, a "good fit" is usually found through nonlinear least-squares minimization. We find that the least-squares cost function is not very sensitive to changes in the FFCF parameters. Therefore, we recommend avoiding this method if the sole purpose is extracting the FFCF—the CLS analysis is far more robust. Unfortunately, both methods are highly susceptible to systematic error. There exists a third approach to extract the FFCF which involves scanning T at a fixed non-zero value of τ for rephasing and non-rephasing spectra [46]. We have not yet tried this method.

3.7 Data Interpretation

The physical meaning of the CLS curve should be understood in terms of the FFCF. This correlation function describes the ensemble of environmental fluctuations as seen by the vibrational label. To link this to environment fluctuations in a purely molecular picture, one needs to know how those fluctuations affect the instantaneous frequency of the vibrational label. Computationally, this is a difficult task and good reviews are available on this topic [47]. It is possible to gain a qualitative picture of what the FFCF means for a specific system. For this, we recommend the experimental design strategies outlined in Subheading 3.5.4.

A qualitative understanding of the CLS curve begins by fitting the CLS curve to a functional form that is representative of the normalized FFCF. A common class of fitting functions is the sum of decaying exponentials [43]:

$$CLS(T) = \sum_i A_i e^{\frac{-t}{\tau_i}} \tag{3}$$

where A_i represents the intensity of frequency fluctuations and is proportional to $\Delta\omega_i^2$, where $\Delta\omega_i$ is the amplitude of the frequency fluctuation, and τ_i represents the correlation time of that fluctuation. Excluded in this functional form is pure dephasing contribution, T_2^* , which should normally be included in $C(T)$ as an additional term, $\delta(T)/T_2^*$, where $\delta(T)$ is the Dirac-delta function.

Not all the parameters extracted from the CLS are reliable. This is because inhomogeneous fluctuations that occur on the same time scale (or faster) as dephasing become "averaged out" (similar to motional narrowing) and become indistinguishable from homogeneous broadening. Alternatively stated, the short time approximation, under which CLS analysis was derived, is broken [43]. This means that any homogeneous contribution or amplitude of a fast inhomogeneous contribution cannot be extracted directly from the CLS curve. The parameters that are reliable include the time constants and relative amplitudes for the exponentials with time constants slower than the dephasing time. Additional steps are required to extract the pure-dephasing component $\left(T_2^*\right)$ and the unnormalized amplitudes of fluctuations $(\Delta\omega_i)$ [43].

The physical interpretation of Eq. (3) can be understood as the additive contribution of several distinct randomly fluctuation processes that induce frequency fluctuations $(\Delta\omega_i)$ with a given time scale (τ_i). For example, hydrogen bond making and breaking with free water may induce a large amplitude-frequency fluctuation (depending on the label) with a relatively fast time scale (e.g. 1 ps). Hydrogen bond making and breaking with confined water, on the other hand, will likely induce slower fluctuations. The spatial amplitude of molecular fluctuations does not necessarily translate directly into the frequency amplitude. For example, fluctuations induced by motions in the acyl chain region will likely have a smaller frequency amplitude because such fluctuations are unable to strongly perturb the oscillator's instantaneous frequency. In addition, the acyl chain region fluctuations are expected to have a long time constant. When the time constant is too slow to be observed within the experimental time frame, the CLS is usually just fit to a constant offset. This offset is known as a static inhomogeneity. Static inhomogeneity means that there is an ensemble of environments that exchange slowly compared to the experimental time frame.

4 Notes

4.1 Deuterium Exchange

We use the following protocol to prepare deuterium exchanged samples:

Step 1: Prepare a batch of the solution in D_2O with a known *mass* of all components. The idea is that a known volume of high-purity D_2O will be added to the lyophilized components later to achieve a target concentration. This solution would normally contain a mixture of all components required for the system. If the lipid membrane requires preparation, such as sonication, extrusion, or thermal cycling, several solutions can be prepared with the individual components. Any additional preparation steps will be done after the deuterium exchange is complete. If required, buffer salts should be prepared separately.

Step 2: The batch solution or solutions should be separated into several aliquots, each of which will correspond to a single 2D IR data set. The exchange should be allowed for several minutes. One should also consider exchanging any strongly absorbing counterions that may be present in the sample. For example, peptides prepared by solid-phase peptide synthesis usually have TFA as a counter ion from the cleavage step. TFA absorbs strongly around 6 μm. If TFA poses a problem, it can be removed by adding deuterium chloride up to a concentration of approximately 1 mM. This will form the volatile free acid which will be removed in the lyophilization step. Counter ions are usually straight forward to monitor via FTIR, if the excess sample is readily available.

Step 3: Lyophilize the aliquots.

Step 4: Reconstitute the aliquots with D_2O and add DCl if necessary. Repeat the "Lyophilization Step" at least once more time to ensure complete H-D exchange or the "Lyophilization Step" and this step at least two more times to ensure complete removal of TFA. We usually store lyophilized individual aliquots frozen in a desiccator jar.

Step 5: When ready for 2D IR and FTIR measurements, the lyophilized aliquots should be reconstituted with a known volume of high isotopic purity D_2O. At this step, individual components may be mixed and prepared as needed. Caution must be taken to avoid contamination with moisture, especially if lipids need to be thermally cycled or sonicated in a water bath.

Note that we usually take the FTIR and 2D IR data on the sample consecutively. In addition, the specific lipid preparation protocol needs to consider the fact that the above procedure is usually done with low sample volumes (e.g. high isotope purity D_2O often comes in 600 μL ampules).

4.2 Sample Loading

We use the following procedure to load the sample into the sample cell. This procedure works best with low volumes of aqueous samples.

Step 1: Thoroughly clean the sample cell. This usually requires rinsing with DI water and optical grade methanol (e.g. Optima). The CaF_2 windows should be carefully rinsed with DI water and then cleaned with optical grade methanol with folded lens tissue in the same manner as any optical component is cleaned.

Step 2: Partially assemble the sample cell once it is dry. The assembly should include the first o-ring that is resistant to the sample solvent, the first CaF_2 window, the Teflon spacer, and the second CaF_2 window.

Step 3: Using a small stand (such as a small screw wrapped in lens tissue), lift the two CaF_2 windows and Teflon spacers so that the second window is above the top edge of the sample holder such that the top window can be easily removed and replaced. Ideally, the spacer should still reside below the top edge.

Step 4: Lift the second window slightly and pipette in the desired sample volume. The act drawing the sample into the pipette and dispensing onto the first window should be done carefully so that no bubbles are formed. Ideally, the dispensed solution should form a single round drop in the center of the first window.

Step 5: Carefully lower the second window so that the solution evenly spreads between the two windows. At this point, bubbles may be present at the edges of the solution. After tightening the sample cell, these bubbles tend to congregate around the center of the windows.

Step 6: To help remove bubbles, we usually use a Teflon spacer that has a single cut. After the second window is completely lowered, lightly press on the edges of the window to help maneuver the bubbles toward the cut. Usually pressing on the edge opposite to the single cut is enough to remove the bubbles.

Step 7: Once the bubbles are removed, add the second o-ring. Finally, complete and tighten the sample cell assembly.

In our experience, the Harrick cells are relatively airtight. Leakages mostly occur from damaged o-rings or damage on the CaF_2 windows at the o-ring contact point. An easy way to test for leaks is to attach an inflated balloon to one of the cell flow ports with the other port blocked (e.g. using a syringe plunger and valve) and to see if the balloon deflates overnight. As an added precaution, we usually wrap the sample cell with parafilm, although this likely has little effect on the air tightness and the H-D exchange rate.

4.3 Rotating Frame Artifact

If one is not careful, the rotating frame can introduce artifacts into the 2D IR spectrum. These artifacts are introduced by two sources: first, the Nyquist limit in discrete Fourier transform; and second, the symmetry requirements for Fourier transforming over real-valued data. The first effect causes artifacts caused by pulse-shaper nonlinearity to fold over into the spectral region of interest (this is known as aliasing). The folded over artifact can be moved away from the spectral region of interest by changing the τ delay spacing. From our experience, this artifact is usually weak. The second artifact can be more severe. It is caused when spectral features are present on the other side of the rotating frame, as compared to the data (e.g. there are spectral features at negative frequencies). These features are then mirrored across the rotating frame frequency into the spectral region of interest (e.g. these spectral features will be mirrored into positive frequencies). There are two ways to fix this artifact. The first way requires complex phase cycling to extract both the real and imaginary components of the signal. Since the Fourier transform is over complex-valued data, the mirror artifact will not be present. The second method is simply to move the rotating frame further away from the spectral region of interest. A general rule of thumb is to keep the rotating frame at least 1 FWHM away from the pulse center. For example, if the pulse is centered around 2000 cm^{-1} (5 μm) and has an FWHM bandwidth of 150 cm^{-1}, the rotating frame should be set to 1850 cm^{-1} or less. We recommend the second method because it also helps move any residual $1/f$ laser intensity noise away from the spectral region of interest (any $1/f$ noise will be most prominent around the new zero frequency, which is the rotating frame).

References

1. Le Sueur AL, Horness RE, Thielges MC (2015) Applications of two-dimensional infrared spectroscopy. Analyst 140:4336–4349. https://doi.org/10.1039/C5AN00558B
2. Shim S-H, Strasfeld DB, Ling YL, Zanni MT (2007) Automated 2D IR spectroscopy using a mid-IR pulse shaper and application of this technology to the human islet amyloid polypeptide. Proc Natl Acad Sci 104:14197–14202. https://doi.org/10.1073/pnas.0700804104
3. Feng Y, Vinogradov I, Ge N-H (2017) General noise suppression scheme with reference detection in heterodyne nonlinear spectroscopy. Opt Express 25:26262. https://doi.org/10.1364/OE.25.026262
4. Flanagan JC, Baiz CR (2019) Site-specific peptide probes detect buried water in a lipid membrane. Biophys J 116:1692–1700. https://doi.org/10.1016/j.bpj.2019.03.002
5. Osborne DG, Dunbar JA, Lapping JG et al (2013) Site-specific measurements of lipid membrane interfacial water dynamics with multidimensional infrared spectroscopy. J Phys Chem B 117:15407–15414. https://doi.org/10.1021/jp4049428
6. Volkov VV, Chelli R, Righini R (2006) Domain formation in lipid bilayers probed by two-dimensional infrared spectroscopy. J Phys Chem B 110:1499–1501. https://doi.org/10.1021/jp056396o
7. Mukherjee P, Kass I, Arkin IT, Zanni MT (2006) Picosecond dynamics of a membrane protein revealed by 2D IR. Proc Natl Acad Sci 103:3528–3533. https://doi.org/10.1073/pnas.0508833103
8. Hamm P, Zanni M (2011) Concepts and methods of 2D infrared spectroscopy. Cambridge University Press, Cambridge
9. Deol SS, Bond PJ, Domene C, Sansom MSP (2004) Lipid-protein interactions of integral membrane proteins: a comparative simulation study. Biophys J 87:3737–3749. https://doi.org/10.1529/biophysj.104.048397
10. Marius P, Zagnoni M, Sandison ME et al (2008) Binding of anionic lipids to at least three nonannular sites on the potassium channel KcsA is required for channel opening. Biophys J 94:1689–1698. https://doi.org/10.1529/biophysj.107.117507
11. Schmidt D, Jiang Q-X, MacKinnon R (2006) Phospholipids and the origin of cationic gating charges in voltage sensors. Nature 444:775–779. https://doi.org/10.1038/nature05416
12. Smith AW (2012) Lipid–protein interactions in biological membranes: a dynamic perspective. Biochim Biophys Acta Biomembr 1818:172–177. https://doi.org/10.1016/j.bbamem.2011.06.015
13. Niu S-L, Mitchell DC, Litman BJ (2001) Optimization of receptor-G protein coupling by bilayer lipid composition II. J Biol Chem 276:42807–42811. https://doi.org/10.1074/jbc.M105778200
14. Hidalgo C, Thomas DD, Ikemoto N (1978) Effect of the lipid environment on protein motion and enzymatic activity of sarcoplasmic reticulum calcium ATPase. J Biol Chem 253:6879–6887
15. Volkov VV, Chelli R, Zhuang W et al (2007) Electrostatic interactions in phospholipid membranes revealed by coherent 2D IR spectroscopy. Proc Natl Acad Sci 104:15323–15327. https://doi.org/10.1073/pnas.0706426104
16. Volkov VV, Takaoka Y, Righini R (2009) What are the sites water occupies at the interface of a phospholipid membrane? J Phys Chem B 113:4119–4124. https://doi.org/10.1021/jp806650c
17. Volkov VV, Chelli R, Muniz-Miranda F, Righini R (2011) Structural properties of a membrane associated anchor dipeptide. J Phys Chem B 115:5294–5303. https://doi.org/10.1021/jp109284z
18. Stevenson P, Tokmakoff A (2017) Time-resolved measurements of an ion channel conformational change driven by a membrane phase transition. Proc Natl Acad Sci 114:10840–10845. https://doi.org/10.1073/pnas.1708070114
19. Stevenson P, Tokmakoff A (2017) Ultrafast fluctuations of high amplitude electric fields in lipid membranes. J Am Chem Soc 139:4743–4752. https://doi.org/10.1021/jacs.6b12412
20. Stevenson P, Tokmakoff A (2018) Infrared insights into the effect of cholesterol on lipid membranes. Chem Phys 512:146–153. https://doi.org/10.1016/j.chemphys.2017.12.012
21. Haidekker MA, Brady T, Wen K et al (2002) Phospholipid-bound molecular rotors:

synthesis and characterization. Bioorg Med Chem 10:3627–3636. https://doi.org/10.1016/S0968-0896(02)00240-7
22. Vinogradov IV (2019) Technological improvements in two-dimensional infrared spectroscopy and their applications to biological systems. Univeristy of California, Irvine
23. Hamm P, Kaindl RA, Stenger J (2000) Noise suppression in femtosecond mid-infrared light sources. Opt Lett 25:1798. https://doi.org/10.1364/OL.25.001798
24. Waegele MM, Culik RM, Gai F (2011) Site-specific spectroscopic reporters of the local electric field, hydration, structure, and dynamics of biomolecules. J Phys Chem Lett 2:2598–2609. https://doi.org/10.1021/jz201161b
25. Kim H, Cho M (2013) Infrared probes for studying the structure and dynamics of biomolecules. Chem Rev 113:5817–5847. https://doi.org/10.1021/cr3005185
26. Ma J, Pazos IM, Zhang W et al (2015) Site-specific infrared probes of proteins. Annu Rev Phys Chem 66:357–377. https://doi.org/10.1146/annurev-physchem-040214-121802
27. Ghosh A, Ostrander JS, Zanni MT (2017) Watching proteins wiggle: mapping structures with two-dimensional infrared spectroscopy. Chem Rev 117:10726–10759. https://doi.org/10.1021/acs.chemrev.6b00582
28. Palfrey SL, Heinz TF (1985) Coherent interactions in pump–probe absorption measurements: the effect of phase gratings. J Opt Soc Am B 2:674. https://doi.org/10.1364/JOSAB.2.000674
29. Mukamel S (1999) Principles of nonlinear optical spectroscopy. Oxford University Press, Oxford
30. Bakulin AA, Cringus D, Pieniazek PA et al (2013) Dynamics of water confined in reversed micelles: multidimensional vibrational spectroscopy study. J Phys Chem B 117:15545–15558. https://doi.org/10.1021/jp405853j
31. Tsai Y-S, Ma S-M, Nishimura S, Ueda I (1990) Infrared spectra of phospholipid membranes: interfacial dehydration by volatile anesthetics and phase transition. Biochim Biophys Acta Biomembr 1022:245–250. https://doi.org/10.1016/0005-2736(90)90120-D
32. Genova J, Petrov M, Bivas I et al (2018) Fourier-transform infrared and Raman characterization of bilayer membranes of the phospholipid SOPC and its mixtures with cholesterol. Colloids Surfaces A Physicochem Eng Asp 557:85–93. https://doi.org/10.1016/j.colsurfa.2018.04.044
33. Barth A (2007) Infrared spectroscopy of proteins. Biochim Biophys Acta Bioenerg 1767:1073–1101. https://doi.org/10.1016/j.bbabio.2007.06.004
34. Kim YS, Hochstrasser RM (2005) Chemical exchange 2D IR of hydrogen-bond making and breaking. Proc Natl Acad Sci USA 102:11185–11190. https://doi.org/10.1073/pnas.0504865102
35. Ross MR, White AM, Yu F et al (2015) Histidine orientation modulates the structure and dynamics of a de novo metalloenzyme active site. J Am Chem Soc 137:10164–10176. https://doi.org/10.1021/jacs.5b02840
36. Kel O, Tamimi A, Fayer MD (2014) Size-dependent ultrafast structural dynamics inside phospholipid vesicle bilayers measured with 2D IR vibrational echoes. Proc Natl Acad Sci 111:918–923. https://doi.org/10.1073/pnas.1323110111
37. Bloem R, Garrett-Roe S, Strzalka H et al (2010) Enhancing signal detection and completely eliminating scattering using quasi-phase-cycling in 2D IR experiments. Opt Express 18:27067. https://doi.org/10.1364/OE.18.027067
38. Shim S-H, Zanni MT (2009) How to turn your pump–probe instrument into a multidimensional spectrometer: 2D IR and Vis spectroscopiesvia pulse shaping. Phys Chem Chem Phys 11:748–761. https://doi.org/10.1039/B813817F
39. Shim S-H, Strasfeld DB, Zanni MT (2006) Generation and characterization of phase and amplitude shaped femtosecond mid-IR pulses. Opt Express 14:13120. https://doi.org/10.1364/OE.14.013120
40. Karthick Kumar SK, Tamimi A, Fayer MD (2012) Comparisons of 2D IR measured spectral diffusion in rotating frames using pulse shaping and in the stationary frame using the standard method. J Chem Phys 137:184201. https://doi.org/10.1063/1.4764470
41. Zanni MT, Gnanakaran S, Stenger J, Hochstrasser RM (2001) Heterodyned two-dimensional infrared spectroscopy of solvent-dependent conformations of acetylproline-NH_2. J Phys Chem B 105:6520–6535. https://doi.org/10.1021/jp0100093
42. Kramer PL, Giammanco CH, Tamimi A et al (2016) Quasi-rotating frame: accurate line shape determination with increased efficiency in noncollinear 2D optical spectroscopy. J Opt Soc Am B 33:1143. https://doi.org/10.1364/JOSAB.33.001143
43. Kwak K, Park S, Finkelstein IJ, Fayer MD (2007) Frequency–frequency correlation

functions and apodization in two-dimensional infrared vibrational echo spectroscopy: a new approach. J Chem Phys 127:124503. https://doi.org/10.1063/1.2772269
44. Kwak K, Rosenfeld DE, Fayer MD (2008) Taking apart the two-dimensional infrared vibrational echo spectra: more information and elimination of distortions. J Chem Phys 128:204505. https://doi.org/10.1063/1.2927906
45. Guo Q, Pagano P, Li Y-L et al (2015) Line shape analysis of two-dimensional infrared spectra. J Chem Phys 142:212427. https://doi.org/10.1063/1.4918350
46. Osborne DG, Kubarych KJ (2013) Rapid and accurate measurement of the frequency–frequency correlation function. J Phys Chem A 117:5891–5898. https://doi.org/10.1021/jp307854f
47. Błasiak B, Londergan CH, Webb LJ, Cho M (2017) Vibrational probes: from small molecule solvatochromism theory and experiments to applications in complex systems. Acc Chem Res 50:968–976. https://doi.org/10.1021/acs.accounts.7b00002

Chapter 12

Estimation and Imaging Techniques to Study Lipids in Mammalian Samples

Sudhanshu Shukla and Sanghamitra Mishra

Abstract

Lipids are small biomolecules that constitute structural elements of the cellular membranes and play crucial functions to maintain various cellular processes indispensable for survival. In general, there are four types of lipids existing in the human body mainly fatty acids, glycerides, non-glycerides, and complex lipids. Abnormal lipid levels and signaling give rise to a number of clinical pathologies and disorders including metabolic diseases, cardiovascular diseases, diabetes, and immune dysfunctions. Some of these, such as sitosterolemia, hypercholesterolemia, fish eye disease, age-related macular degeneration, Tangier disease, etc., are inherited and caused by mutations in genes from lipid pathways. Therefore, it is important to study the molecular mechanisms of disease progression and establishing correlation between lipid levels and grade of clinical onset to understand the pathophysiologies of various lipid-associated diseases. This requires reliable imaging and quantification techniques of various lipids. In this chapter, by taking samples from human eyes and cell culture models, we have explained how imaging and quantification techniques can provide molecular details toward understanding the correlation between lipids and eye disease. We describe the use of standard kits and techniques available, and some of these protocols can be extrapolated to other organ systems with specific modifications.

Keywords Membrane lipids, Cholesterol, LDL, FM dyes, Filipin

1 Introduction

Lipids are essential constituents of cells and tissues. Defects in lipid synthesis, intracellular and extracellular transport, and turnover are a cause of several diseases [1]. Recent work on lipid metabolism has also been linked to chromatin regulation in aging [2]. Among lipids, the role of cholesterol including aberrant ratio of HDL and LDL has been very well linked with various diseases such as diabetes, coronary artery diseases, and other metabolic disorders [3]. Among human organ systems, eye is a complex organ with multiple tissues. Growing evidence suggests that the clinical onset of age-related macular degeneration is catalyzed by cholesterol deposition [4]. Therefore, identification and quantification of various lipids in clinical samples is highly desirable to understand the

Rajendra Prasad and Ashutosh Singh (eds.), *Analysis of Membrane Lipids*, Springer Protocols Handbooks,
https://doi.org/10.1007/978-1-0716-0631-5_12,

disease and its grade. Additionally, reliable quantification and identification technologies are also needed for research purpose to understand the molecular mechanism of the progression of various diseases. Current technologies involve mainly nuclear magnetic resonance, gas chromatography–mass spectrometry, and few biochemical assays for this purpose [5]. Thin-layer chromatography is also widely used in certain scenarios [6]. Imaging techniques involving fluorescent dyes to identify and locate the position either inside the cells or in tissue samples are crucial aspects for a better understanding of related disease biology. Imaging techniques should be used in combination with other lipid identification and quantification approaches to get the in-depth details of the studied diseases. Therefore, in this chapter, we provide details on current approaches and methods for imaging and quantification of cholesterol and LDL by exploiting cell culture model and human eye samples.

2 Staining of Membrane Lipids

FM dye stock solution:FM 4-64-100 μg + 20 μl DMSO = 5 μg/μl (1000×).

FM dye working solution:

12 μl of 5 μg/μl FM dye + 12 ml HBSS-Ca^{++}-Mg^{++} = 5 μg/ml.

Filipin III stock solution:

Filipin 1 mg dissolved in 1 ml 100% ethanol (1 mg/ml) + 9 ml PBS in black tube = 100 μg/ml aliquots stored in 0.5 ml black tubes at −80 °C. Alternatively, 1 mg filipin was dissolved in 200 μl ethanol, aliquoted at 20 μl in 1.5 ml black tubes, and desiccated under argon. Each aliquot had ~100 μg filipin diluted to 0.05 mg/ml in buffer as per subsequent protocol.

ICC buffer:

0.5% Bovine serum albumin, 0.2% Tween 20, 0.05% sodium azide in 1× PBS, pH 7.3.

Phospholipids, glycolipids, and cholesterol are three major classes of membrane lipids, which form the lipid bilayer of cells and regulate several cellular functions. The eye is a complex organ with multiple tissues, viz., cornea, lens, and multilayered retina. Membranes and their constituent lipids play an important role in the functioning of these structures. In fact, cholesterol homeostasis is critical to the many ocular functions, for example, the role of cholesterol in photoreceptor shedding [7]. The lens cell membrane is highly rich in cholesterol [8, 9]. These cells can also synthesize, degrade, and remodel lipids, and the lipid composition determines membrane fluidity helpful in functioning of lens [10, 11].

2.1 Staining With FM Dyes

Lipophilic styryl dyes, commonly known as FM dyes, fluoresce upon binding to membrane lipids and are therefore useful to study membrane activities such as vesiculation. It is used in different model systems ranging from plants to mammalian cells and tissues [12]. One of the primary areas of use of FM dye staining has been to study exocytosis, endocytosis, and synaptic vesicle recycling in neurons.

FM dyes are nontoxic to cells and therefore can be applied with practically no technique-induced cytotoxicity and minimal effect to the treatment being studied. These are practically nonfluorescent in aqueous media and magically fluoresce when they interact/bind to membrane lipids like the outer leaflet of the plasma membrane. These are available in different fluorescent ranges from Molecular Probes, Thermo Fisher [13, 14]. There are also versions of the dyes which use aliphatic additives that can crosslink with glutaraldehyde or formaldehyde. These fixable versions are useful to retain fluorescence while imaging. The hydrophilic version of FM dyes is useful in quantitative assays for faster destaining. Here we describe a method to stain the plasma membrane of cultured human retinal pigment epithelial cells. This method can be adapted to other cell culture systems.

2.2 FM Dye Staining for Cells Grown on Transwells or Chambered Slides

Transwell permeable supports are removable cell culture supports that can be used for anchorage-dependent and anchorage-independent cell types. It is a useful tool for researchers to study cell migration and intracellular processes. They are available in different pore sizes, materials, like polyester, polycarbonate, or PTFE, and coatings, viz., laminin, collagen, etc.

Cells are also grown on glass slides or coverslips for imaging experiments. Chambered slides are useful when there are multiple treatments or cell types have to be imaged.

FM dye used: FM4-64FX.

Working staining solution: 5 μg/mL dye in ice-cold HBSS. This working solution is to be kept ice cold.

Note: *The presence of divalent cations speeds up endocytosis of the dye, resulting in staining of internal vesicles and membrane structures. Therefore, while staining plasma membrane, buffers and media must be Ca^{++} and Mg^{++} free. While studying endocytosis, this caution can be avoided.*

- Remove medium from wells, wash once with Ca^{++}- and Mg^{++}-free HBSS, and add ice-cold FM dye solution on ice, for 1 min.

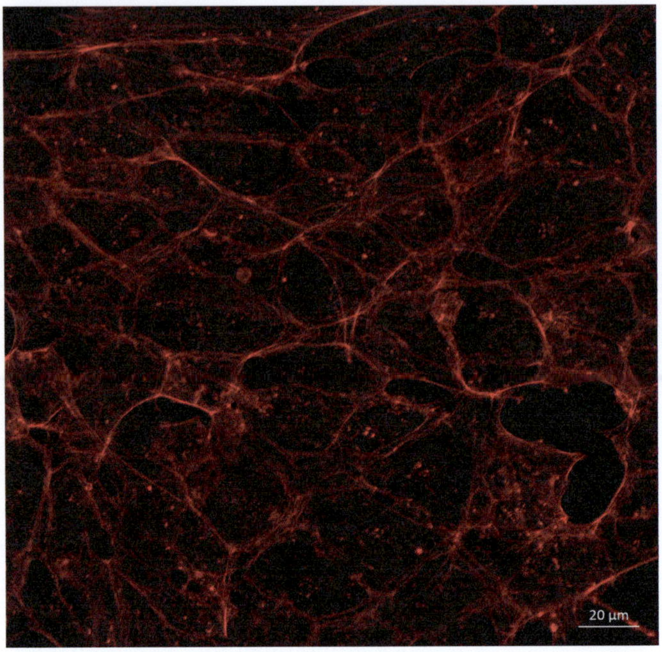

Fig. 1 ARPE-19 cells labeled with FM4-64FX dye. Plasma membrane and some internal vesicle membranes are shown in red

For transwells, washing and staining solution has to be added on both sides of the wells. The plasma membranes will stain quickly.

- Do not wash.
- Fix the stain with ice-cold 4% formaldehyde in Ca^{++}- and Mg^{++}-free HBSS; place on ice for 10 min.
- Rinse the coverslip three times with Ca^{++}- and Mg^{++}-free HBSS.
- Mount on to slides and visualize under a fluorescence or confocal microscope (Fig. 1).

3 Cholesterol

Cholesterol is popularly known as a waxy deposit in blood vessels causing disease, though it is found in all the cells in our bodies and is important for their normal functioning by facilitating the transport of lipid-soluble molecules or making essential molecules such as hormones. It is found in foods and is also synthesized in the liver. Intracellular cholesterol is present as esterified cholesterol for transport to or as unesterified or "free" form. Under certain stress conditions such as starvation, intracellular levels of unesterified cholesterol increase.

Filipin III is derived from *Streptomyces filipinensis*, is a lipophilic polyene fluorescent probe, and is also an antibiotic. Its antibiotic properties involve altering membrane permeability and associated functions by binding to sterols in fungal membranes and inhibiting the conversion of the normal membrane-bound prion protein (PrP-sen) to its pathological isoform (PrP-res), which is a key event in pathogenesis.

Oil Red O, on the other hand, is a lipid-soluble dye, which is used to detect esterified cholesterol. Filipin and Oil Red O have been used to simultaneously detect and differentiate free and esterified cholesterol [15]. Filipin stains free cholesterol but does not stain cholesteryl ester and because Oil Red O stains cholesteryl ester but does not stain free cholesterol.

Here we refer to its use in cholesterol labeling. It can bind to unesterified cholesterol and fluoresce in the UV range, excitation: 405 nm and emission: 410–500 nm. It is used widely to study localization of unesterified cholesterol by imaging or quantitate free cholesterol by measuring the complex in UV range. Filipin is also used for the detection of lipoproteins on immobilized membranes and polyacrylamide gels.

3.1 Detecting Endogenous Unesterified Cholesterol by Imaging

3.1.1 Filipin Staining

Note:

- Filipin is extremely light sensitive. Small aliquots of prepared stock must be saved in light-protected black tubes at −80 °C.
- All processes from addition of the reagent onward must be done in dark.
- Filipin stained samples must be immediately assayed for quantitative results.
- Storage in dark may retain fluorescence but will not be representative of the exact amount of unesterified cholesterol while comparing samples.
- Co-staining with nuclear stain DAPI cannot be done as both DAPI and filipin fluoresce in the UV range. Nuclear stains with emissions in other wavelengths such as Topro-3 (far red) from Molecular Probes can be used. This also applies to mounting media with added nuclear stains.
- Filipin photobleaches very fast under laser especially at high magnifications (>40×). Therefore, quantitative comparisons between treatments at these fields must be done carefully with multiple replicate fields.
- Filipin staining works with both phosphate buffer saline and Tris buffer saline. If co-staining with antibodies, TBS can be used.

There are kits available for cholesterol detection in cells such as the Cell-Based Detection Assay Kit from Cayman Chemical. Making your own dilutions is a more economical alternative, which is described below.

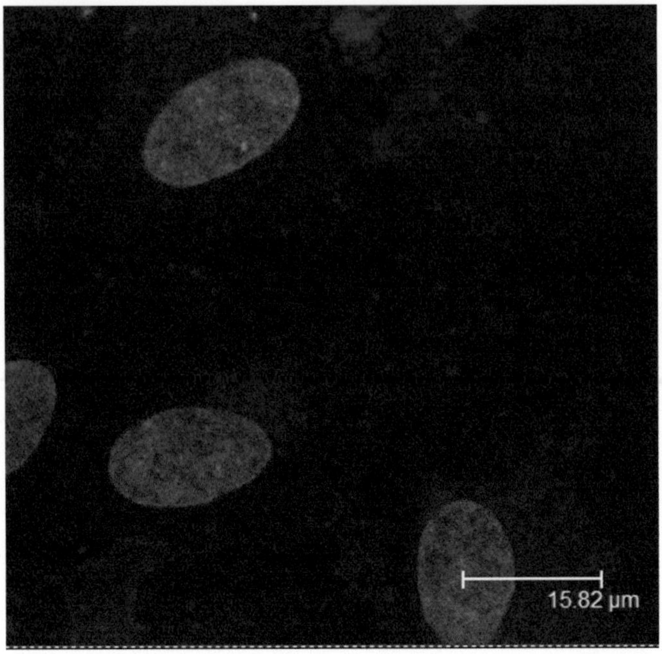

Fig. 2 Confocal image of ARPE-19 cells labeled with filipin and Topro-3 showing intracellular unesterified cholesterol (blue) and nuclei (gray)

Filipin working solution: 100 µg/ml filipin (prepared fresh) added and incubated in dark for 2 h.

Wash with 1× phosphate buffer saline and fix with 4% paraformaldehyde treated cells grown on glass slides for 10–15 min at RT.

Wash fixed cells once with TBS and permeabilize with 0.1% Polysorbate 20 (Tween 20) and twice with TBS for 5 min each (this step is required if co-staining with antibodies). Filipin itself does not require permeabilization.

Add diluted filipin to each well and incubate in the dark for 60 min.

Following filipin labeling, cells can be washed and labeled with nuclear stain, which fluoresces other than UV range such as Topro-3 (Molecular Probes).

After that, wash cells in dark three times with TBS and mount with mounting agent such as Prolong Gold (Molecular Probes).

Staining can be immediately visualized using a microscope equipped with UV filters, excitation: 340–380 nm, and emission: 385–470 nm such as Olympus FV1000 confocal microscope (Fig. 2).

3.1.2 Flat Membrane Immunostaining with Filipin

This method is for staining cells grown on transwells. Membranes are fixed with 4% PFA in PBS with intact transwell followed by incubation with 0.1 M glycine (to stop PFA action) for 10 min at RT.

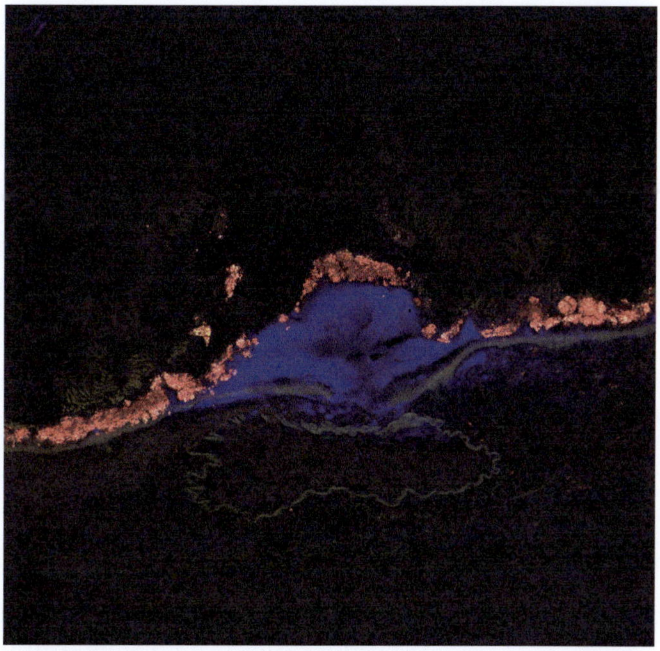

Fig. 3 Filipin labeling in a frozen postmortem human eye section showing a druse filled with unesterified cholesterol (blue). The autofluorescence of lipofuscin filled overlying retinal pigment epithelium appears as brick colored. Picture courtesy Dr. Campos Maria NEI, Bethesda

Cells were washed once with PBS and once for 7 min at RT with ICC buffer to permeabilize for phalloidin, nuclear, or immunostaining. Note that any labeling following the addition of filipin must be performed in dark. Filipin staining can also be performed in tissue sections, Fig. 3 [16].

3.2 Oil Red O Staining

Oil Red O stains intracellular lipid materials in red-orange color. Intracellular lipid droplets and lipid-containing vacuoles in cells can be identified in cell culture models or histological sections. One limitation is that Oil Red O staining can be done on fresh samples and not fixed samples as lipids get depleted during the fixation process [17].

3.3 LDL Uptake Assay

Low-density lipoproteins (LDLs) are the major carrier of cholesterol in the blood, accounting for more than 60% of total plasma cholesterol. LDL is taken up by hepatic and extrahepatic tissues through LDL receptor-mediated endocytosis where it is internalized and degraded to free cholesterol and amino acids. Uptake of native and oxidized LDL by monocyte-derived macrophages in the vessel wall results in accumulation of cholesterol and formation of atherosclerotic plaques. The mechanism underlying the uptake of LDL by both hepatic tissue and cells of the cardiovascular system is not clearly understood. Thus, LDL uptake and its regulation are important therapeutic targets for atherosclerosis and related diseases. Cayman's LDL Uptake Cell-Based Assay Kit provides a convenient tool for studying LDL uptake and regulation at cellular level. The kit employs human LDL conjugated to DyLight™ 550 as a fluorescent probe for detection of LDL uptake into cultured cells. An LDL receptor-specific polyclonal antibody and a DyLight™ 488-conjugated secondary antibody are included for identifying the distribution of LDL receptors.

3.3.1 Tracking LDL Uptake by Labeled LDL Imaging

LDL conjugated to DyLight 549 is diluted 1:100 (500 μl each well) in sterile-filtered colorless serum free medium (SFM).

After treatment, wash the cells once with colorless SFM followed by incubation in a 1:100 solution of LDL DyLight 549 working solution in SFM for 3 h.

Followed by washing twice with colorless SFM to remove external LDL-DyLight, add colorless SFM medium into all wells followed by imaging the "live cells" for "LDL uptake."

Alternatively, to preserve labeled LDL uptake, cells (grown on coverslips), after live imaging, can be washed with PBS (137 mM NaCl, 10 mM phosphate, 2.7 mM KCl; pH: 7.4) and fixed in 4% paraformaldehyde in PBS for 10 min, mounted and visualized using a Zeiss LSM 700 confocal microscope.

After imaging for LDL uptake, cells can be fixed and proceeded for immunolabeling protocols.

Briefly, cells are washed in TBS, fixed in 4% paraformaldehyde for 10 min at RT, washed in TBST, blocked in kit-provided blocking solution, and incubated with primary antibody O/N at 4 °C.

This is followed by two washes with TBST and incubation with secondary antibody (conjugated to DyLight 488, fluorescing at wavelengths other than DyLight 549-red range), followed by three washes in TBST, mounting and imaging.

Notes: *For protocols that involve fixing of cells, grow cells on glass coverslips.*

Nonspecific labeling can be checked by cells with unlabeled LDL prior to incubation with DyLight conjugated LDL uptake [18].

3.4 Quantitation of Cholesterol

Quantitation of intracellular cholesterol or cholesterol in plasma, cell media, or other body fluids can be done by plate-based fluorometric measurement [18, 19]. This estimation relies on the measurement of hydrogen peroxide during cholesterol oxidation. Cholesterol is formed by hydrolysis of cholesteryl esters followed by its oxidation by cholesterol oxidases resulting in ketones and hydrogen peroxide. The peroxide produced in this process can be estimated by using a stable probe, 10-acetyl-3,7,-dihydroxyphenoxazine (ADHP) which reacts with peroxide to produce fluorescent resorufin which can be measured to estimate cholesterol.

Some of the kits available for quantitation of cholesterol are: Cholesterol Fluorometric Assay Kit, Cayman Chemical, and Cholesterol/Cholesteryl Ester Quantification Kit from Abcam Inc.

Additionally, image quantification after imaging or direct fluorescence quantification on confocal can quantify differences in the lipids assayed for between treatments.

4 Conclusion

Mass spectrometry-based techniques like GC/MS and imaging of lipids using exogenous fluorescent agents for identification and relative quantification of array of lipid species have been proven to be useful in clinical and basic research. Nonetheless, considering the complex nature of various lipids and their associated roles in numbers of diseases, we need more precise and specific tools to quantitate along with microscopic identification of lipid of interest to investigate the intracellular localization in clinical tissue samples as well as in cell cultures. Recent emerging techniques such as "coherent Raman scattering" (CRS) for quantitative imaging to study lipid content in single cells are gaining significant attention [20]. In a similar way, a three-dimensional (3D) label-free imaging and quantification approach for lipid droplets has been utilized in live hepatocytes [21]. Interestingly, in another study, the authors have mapped the lipid distributions quantitatively through visualization by employing magnetic resonance transfer (MRI) in- and opposed-phase (IOP) sequence in gliomas [22]. Therefore, other recent approaches should also be considered along with methods described here in this chapter for detailed understanding.

Acknowledgments

The images shown in this chapter were part of work done in Dr. Graeme J. Wistow's lab, supported by Intramural Program of the National Eye Institute. We thank Dr. Graeme J. Wistow, Dr. Katherine Peterson, Dr. Dinusha Rajapakse, Dr. Campos Maria, and Biological Imaging Core of the National Eye Institute Bethesda for kindly sharing the images.

References

1. From the editors (2008) Nat Rev Mol Cell Biol 9:89. https://doi.org/10.1038/nrm2344
2. Papsdorf K, Brunet A (2019) Linking lipid metabolism to chromatin regulation in aging. Trends Cell Biol 29(2):97–116
3. Lin CJ, Lai CK, Kao MC, Wu LT, Lo UG, Lin LC, Chen YA, Lin H, Hsieh JT, Lai CH, Lin CD (2015) Impact of cholesterol on disease progression. Biomedicine (Taipei) 5(2):7
4. Fan Q, Maranville JC, Fritsche L, Sim X, Cheung CMG, Chen LJ, Gorski M, Yamashiro K, Ahn J, Laude A, Dorajoo R, Lim TH, Teo YY, Blaustein RO, Yoshimura N, Park KH, Pang CP, Tai ES,

Khor CC, Wong TY, Runz H, Cheng CY (2017) HDL-cholesterol levels and risk of age-related macular degeneration: a multiethnic genetic study using Mendelian randomization. Int J Epidemiol 46(6):1891–1902
5. Khoury S, Canlet C, Lacroix MZ, Berdeaux O, Jouhet J, Bertrand-Michel J (2018) Quantification of lipids: model, reality, and compromise. Biomolecules 8(4):E174
6. Fuchs B, Süss R, Teuber K, Eibisch M, Schiller J (2010) Lipid analysis by thin-layer chromatography—a review of the current state. J Chromatogr A 1218(19):2754–2774
7. Albert AD, Boesze-Battaglia K (2005) The role of cholesterol in rod outer segment membranes. Prog Lipid Res 44(2–3):99–124
8. Fleschner CR, Cenedella RJ (1991) Lipid composition of lens plasma membrane fractions enriched in fiber junctions. J Lipid Res 32 (1):45–53
9. Gülcan HG, Alvarez RA, Maude MB, Anderson RE (1993) Lipids of human retina, retinal pigment epithelium, and Bruch's membrane/choroid: comparison of macular and peripheral regions. Invest Ophthalmol Vis Sci 34 (11):3187–3193
10. Zelenka PS (1984) Lens lipids. Curr Eye Res 3 (11):1337–1359
11. Borchman D, Yappert MC (2010) Lipids and the ocular lens. J Lipid Res 51(9):2473–2488
12. Jelínková A, Malínská K, Petrášek J (2019) Using FM dyes to study endomembranes and their dynamics in plants and cell suspensions. Methods Mol Biol 1992:173–187
13. https://www.thermofisher.com/search/results?query=FM+dyes&persona=DocSupport&navId=0&resultPage=1&resultsPerPage=15&refinementAction=true&focusarea=Search+All&sort=relevancy
14. https://www.thermofisher.com/search/results?query=FM+dyes&persona=DocSupport&navId=4294959596&refinementAction=true&focusarea=Search+All&docTypes=
15. Kruth HS, Fry DL (1984) Histochemical detection and differentiation of free and esterified cholesterol in swine atherosclerosis using filipin. Exp Mol Pathol 40(3):288–294
16. Wyatt MK, Tsai JY, Mishra S, Campos M, Jaworski C, Fariss RN, Bernstein SL, Wistow G (2013) Interaction of complement factor h and fibulin3 in age-related macular degeneration. PLoS One 8(6):e68088
17. Rajapakse D, Peterson K, Mishra S, Wistow G (2017) Serum starvation of ARPE-19 changes the cellular distribution of cholesterol and Fibulin3 in patterns reminiscent of age-related macular degeneration. Exp Cell Res 361 (2):333–341
18. Mishra S, Peterson K, Yin L, Berger A, Fan J, Wistow G (2016) Accumulation of cholesterol and increased demand for zinc in serum-deprived RPE cells. Mol Vis 22:1387–1404
19. https://www.caymanchem.com/pdfs/10007640.pdf
20. Gupta A, Dorlhiac GF, Streets AM (2019) Quantitative imaging of lipid droplets in single cells. Analyst 144(3):753–765
21. Seow P, Narayanan V, Hernowo AT, Wong JHD, Ramli N (2018) Quantification and visualization of lipid landscape in glioma using in -and opposed-phase imaging. Neuroimage Clin 20:531–536
22. Kim K, Lee S, Yoon J, Heo J, Choi C, Park Y (2016) Three-dimensional label-free imaging and quantification of lipid droplets in live hepatocytes. Sci Rep 6:36815

MIX
Papier aus verantwortungsvollen Quellen
Paper from responsible sources
FSC® C105338

If you have any concerns about our products,
you can contact us on
ProductSafety@springernature.com

In case Publisher is established outside the EU,
the EU authorized representative is:
Springer Nature Customer Service Center GmbH
Europaplatz 3, 69115 Heidelberg, Germany

Printed by Libri Plureos GmbH
in Hamburg, Germany